STEPHEN WOLFRAM

D1309594

ADDENDUM TO
THE *MATHEMATICA*® BOOK

WOLFRAM
MEDIA

This addendum updates:

The Mathematica Book, 3rd ed.
(Wolfram Media/Cambridge University Press, 1996)
ISBN 0–9650532–0–2/0–521–58889–8 (hardback)
ISBN 0–9650532–1–0/0–521–58888–X (paperback)

(A catalog record for The Mathematica Book is available as LC 96–7218)

**A fully updated, complete edition of
The Mathematica Book will be available as:**

The Mathematica Book, 4th ed.
(Wolfram Media/Cambridge University Press, 1999)
ISBN 1–57955–004–5/0–521–64314–7

Further copies of this addendum are available at:
www.wolfram.com/bookstore

**Complete versions of *The Mathematica Book* are
available from:**
www.wolfram.com/bookstore
or your local bookseller

**Comments on *The Mathematica Book* will be
welcomed at:**
comments@wolfram.com

Bibliographic information for this document:
Stephen Wolfram,
The Mathematica Book, 3rd. ed. Addendum.
(Wolfram Media, 1999)
ISBN 1–57955–012–6

WOLFRAM
MEDIA

Wolfram Media, Inc.
web: www.wolfram–media.com; *email:* info@wolfram–media.com
phone: +1–217–398–9090; *fax:* +1–217–398–9095
mail: 100 Trade Center Drive, Champaign, IL 61820, USA

WOLFRAM RESEARCH
www.wolfram.com

Wolfram Research, Inc.
web: www.wolfram.com
email: info@wolfram.com
phone: 217–398–0700
fax: 217–398–0747
mail: 100 Trade Center Drive
　　　Champaign, IL 61820–7237
　　　USA

Wolfram Research Europe Ltd.
web: www.wolfram.co.uk
email: info@wolfram.co.uk
phone: +44–(0)1993–883400
fax: +44–(0)1993–883800
mail: 10 Blenheim Office Park
　　　Lower Road, Long Hanborough
　　　Oxfordshire OX8 8LN
　　　UNITED KINGDOM

Wolfram Research Asia Ltd.
web: www.wolfram.co.jp
email: info@wolfram.co.jp
phone: +81–(0)3–5276–0506
fax: +81–(0)3–5276–0509
mail: Izumi Building 8F
　　　3–2–15 Misaki-cho
　　　Chiyoda-ku, Tokyo 101–0061
　　　JAPAN

MATHEMATICA®

Note: All services are also available at wolfram.co.uk *and* wolfram.co.jp, *as well as at* wolfram.com.

General information:
info@wolfram.com
www.wolfram.com

**Frequently asked technical
and other questions:**
www.wolfram.com/faq

**User registration and password
requests:**
register@wolfram.com
www.wolfram.com/register
*Institutional and other non-owner users
are also encouraged to register.*

Technical support and bug reports:
support@wolfram.com
www.wolfram.com/support
Support is available only to registered users.

***Mathematica* products:**
orders@wolfram.com
www.wolfram.com/orders
*Order all Wolfram Research products
and upgrades online.*

***Mathematica* books and miscellanea:**
www.wolfram.com/bookstore

***MathSource* Electronic Library:**
www.mathsource.com
ftp.mathsource.com

MathWire electronic newsletter:
www.wolfram.com/mathwire

mathgroup newsgroup:
news: comp.soft-sys.math.mathematica
www.wolfram.com/mathgroup

Suggestions:
suggest@wolfram.com
www.wolfram.com/suggestions

***Mathematica* Archive:**
*Wolfram Research maintains an archive
of Mathematica-related documents.
Publications and other non-proprietary
material are welcome at:*

　　　The *Mathematica* Archive
　　　Wolfram Research, Inc.
　　　100 Trade Center Drive
　　　Champaign, IL 61820–7237, USA
　　　email: archive@wolfram.com
　　　web: www.wolfram.com/archive

Some New Features of *Mathematica* 4

Mathematica 4 introduces important extensions to the *Mathematica* system, especially major efficiency enhancements in handling large volumes of numerical data. It also includes a range of new algorithmic, language and interface features. Throughout this book, items that are new in 4 are indicated by ⁺■ ; those that were changed are indicated by ˷■ . Except as noted in Section A.13, 4 is fully compatible with all earlier versions.

Numerical Computation

- Internal packed array technology to make repetitive operations on large numerical datasets radically more efficient in speed and memory.

- Highly optimized algorithms for doing computations on numbers with up to millions of digits.

- Faster input and output of very large integers.

- Complete preservation of precision on input and output of approximate real numbers.

- Convolution and correlation of arrays of any dimension.

- New optimized algorithms for Fourier transforms.

- Faster solution of numerical polynomial equations.

- New algorithms for FindMinimum.

- Direct support for matrix traces.

Algebraic Computation

- Support for assumptions in Simplify, FunctionExpand and related functions.

- Specification of domains for variables.

- Many additional transformations in FullSimplify and FunctionExpand.

- Simplification of polynomial and other inequalities.

- Full support of symbolic Laplace, Fourier and Z transforms.

- Extensions to integration and summation.

- Extensions to transcendental equation solving.

- Faster multiple differentiation.

- Support for subresultants.

Mathematical Functions

- Dirac delta and other generalized functions.

- Struve functions.

- Nielsen generalized polylogarithms.

- Appell F1 bivariate hypergeometric function.

- Harmonic numbers.

- Khinchin and Glaisher constants.

- Multiplicative order and Carmichael lambda functions.

- New optimized methods for evaluating e, π and other constants to very high precision.

- Full support for continued fractions.

- Support for periodic digit sequences.

- Direct support for bitwise operations.

Graphics and Sound

- Export of graphics and sound in many formats.

- Import of graphics and sound in many formats.

- Support for discrete scaling of color levels.

- Fully consistent support for absolute options.

- Experimental support for real-time 3D graphics (Windows only).

Programming and Core System

- NestWhile and NestWhileList, allowing generalizations of FixedPoint.

- PadLeft and PadRight.

- Support for padding and overhangs in Partition.

- ListConvolve and ListCorrelate.

- Generalization of Take, Drop and related functions to any number of dimensions and any stride.

- Support for All as a specification of parts at specific levels in expressions.

- Extension to Mod to support cyclic lists.

- `Developer`` context containing advanced and algorithmic specific built-in functions.
- `Experimental`` context providing a preview of features under development.

Input and Output

- Optimized minimal-change line breaking for smooth input of expressions and programs.
- Event-oriented cursor tracker for visual continuity during input.
- Dynamic color cues for delimiter matching during input.
- Automatic replacement of input key sequences by special characters or other objects.
- e and i used by default in standard output.
- New optional syntax for part extraction and function application.
- Inline cells within typeset *Mathematica* expressions.
- Much faster string-oriented output of *Mathematica* expressions.

Notebook

- Full-function spell checking including special technical dictionaries.
- Dictionary-validated algorithmic hyphenation.
- Additional keyboard navigation features.

- Platform-independent double buffering to eliminate flicker.
- Optimized controller for smooth autoscrolling.

System Interface

- Streamlined import and export of tabular data.
- Extended support for conversion to HTML.
- Additional support for TeX output.
- Faster *MathLink* external program communication.

Add-ons and Experimental Features

- Direct support for sparse linear algebra.
- Experimental support of quantifier elimination using cylindrical algebraic decomposition.
- Experimental support for symbolic optimization.
- Experimental support for real-time value displays.
- Experimental support for *MathLink*-based remote file systems.
- Experimental support for pop-up palettes.

Some New Features of *Mathematica* 3.0

The first major new version of *Mathematica* in several years, 3.0 strengthens the core computational capabilities of *Mathematica*, and adds some revolutionary new features. Throughout this book, items that are new are indicated by +■ ; those that are changed are indicated by ~■ . Except as noted in Section A.13, 3.0 is fully compatible with earlier versions.

Numerical Computation

- Adaptive precision control to generate results with guaranteed precision.
- High-performance compilation of list-oriented, procedural and functional numerical operations.
- Optimized algorithms for one- and higher-dimensional interpolation.
- Optimized algorithms for solution of differential equations.
- Solution of boundary value ordinary differential equations, and initial value partial differential equations.
- High-dimensional numerical integration.
- Optimized minimization algorithms.
- LU and Jordan decomposition of matrices.
- Numerical differentiation.
- Automatic comparison and manipulation of exact numeric quantities.
- Support for exact implicitly defined algebraic numbers.
- Support for interval arithmetic.
- Fully adjustable global numerical precision control model.
- New capabilities for extracting segments of digits in exact and inexact numbers.
- Machine-independent mechanisms for input and output of numbers without loss of precision.

Algebraic Computation

- Enhanced and optimized simplification of algebraic expressions.
- Simplification of expressions involving special functions.
- Built-in functions for transformations on trigonometric expressions.
- Greatly extended indefinite and definite symbolic integration.

- Support for principal values and assumptions on parameters in integrals.
- Greatly extended symbolic sums and products.
- Greatly extended symbolic solution of ordinary and partial differential equations.
- Optimized symbolic linear algebra.
- Enhanced handling of exact numerical quantities.
- Generation and system-wide support for algebraic numbers.
- Highly optimized Gröbner basis reduction.

Mathematical Functions

- Faster evaluation of many special functions.
- Fresnel integrals and hyperbolic sine and cosine integrals.
- Inverse error function, gamma function and beta function.
- Product log function.
- $_pF_q$ generalized hypergeometric functions and Meijer G functions.
- Additional Weierstrass, elliptic and related functions.
- Mathieu functions.
- Stieltjes constants.
- Built-in Fibonacci numbers and polynomials.

Graphics

- Full typesetting capabilities for labels and text in plots.
- Automatic conversions to EPS, TIFF, GIF and other formats.
- Kernel control of animation in notebooks.
- Absolute offset specifications in graphics primitives.
- Direct control of final graphics size, resolution and so on.
- Direct generation of text strings for graphics in various formats.

Programming and Core System

- Faster execution speed and lower memory usage for typical kernel operations.
- Dumping of function definitions for optimized loading.

- Powerful new general symbolic programming functions, including ReplaceList and Split.

- Pattern-based non-local control flow with Throw and Catch.

- Separate support for verbatim and held patterns.

- Enhanced control of basic evaluation.

- New functions and enhancements for string manipulation.

- Extensive support for manipulation of non-ASCII characters.

Input and Output

- Support for WYSIWYG fully editable two-dimensional typeset input and output.

- Extended *Mathematica* language incorporating special characters and two-dimensional notation.

- Over 700 special characters for mathematical and other notation.

- Support for traditional math textbook notation for output and heuristic input.

- Top-quality typeset output with advanced adjustable layout rules.

- Complete symbolic language for specifying two-dimensional typeset structures.

- Two-dimensional input and manipulation of arrays and matrices.

- Complete support for international character sets and Unicode.

- TEX conversion with line-breaking information.

- Optimized textual import and export of typeset structures.

Notebook Interface

- Programmable documents based on underlying symbolic representation.

- Symbolic language for specifying user interface operations.

- Customizable palettes that can execute any kernel or front end operation.

- Integrated active elements and hyperlinks in notebooks.

- Separate style environments for screen and printing.

- New style sheets for varied document types.

- Language-based control of all features of text, graphics, cells and notebooks.

- Inline typesetting and graphics embedded in text.

- Enhanced text formatting capabilities including full text justification.

- Integrated customizable notebook-based online help with hyperlinks and the full text of this book.

- Fully platform-independent notebook file format.

- Notebook conversion to TEX, HTML and other external formats.

- Option inspector for interactive control of all notebooks and front end properties.

- Keyboard commands for editing and notebook navigation; drag and drop.

System Interface

- Additional kernel directory and file manipulation functions.

- Support for multiplatform external program clusters.

- Optimized *MathLink* external program interface.

- Loopback links for storing expressions in external programs.

- Direct support for arrays in external programs.

- Shared libraries for *MathLink*.

- Uniform layout of system files across all platforms.

- Automatic initialization of kernel and front end functionality as well as documentation for add-ons.

- Support for multilingual versions and all standard keyboard character encodings.

- Direct CD-ROM executability.

- TCP-based network license server.

- OLE support under Microsoft Windows.

Standard Add-on Packages

- Manipulating and solving algebraic inequalities.

- Symmetric polynomials.

- Manipulating quaternions and elements of Galois fields.

- Complete integrals and differential invariants of nonlinear PDEs.

- Z transforms.

- Primitive roots for arbitrary algebraic extensions.

- Numerical residue and series computations.

- Data smoothing and filtering.

- Classical and robust multivariate descriptive statistics.

- Linear and nonlinear regression with diagnostics.

- Simplified arithmetic and algebra without complex numbers.

- Full online documentation of all packages.

Addendum Table of Contents

including sections new or substantially modified for Version 4

~■ substantially modified for Version 4
+■ new for Version 4

Note: Vertical ellipses are used to indicate material unchanged in Version 4 that has been omitted from this addendum.

In this addendum "the complete *Mathematica Book*" refers to *The Mathematica Book*, Third Edition (for *Mathematica* 3.0).

1.1 Numerical Calculations

■ 1.1.5 Complex Numbers

You can enter complex numbers in *Mathematica* just by including the constant I, equal to $\sqrt{-1}$. Make sure that you type a capital I.

If you are using notebooks, you can also enter I as i by typing ⎀ESC⎀ i i ⎀ESC⎀ (see page 36 of the complete *Mathematica Book*). The form i is normally what is used in output. Note that an ordinary i means a variable named i, not $\sqrt{-1}$.

This gives the imaginary number result $2i$.	*In[1]:=* **Sqrt[-4]** *Out[1]=* $2\,i$
This gives the ratio of two complex numbers.	*In[2]:=* **(4 + 3 I) / (2 - I)** *Out[2]=* $1 + 2\,i$
Here is the numerical value of a complex exponential.	*In[3]:=* **Exp[2 + 9 I] //N** *Out[3]=* $-6.73239 + 3.04517\,i$

x + I y	the complex number $x + i\,y$
Re[z]	real part
Im[z]	imaginary part
Conjugate[z]	complex conjugate z^* or $\bar{z}$
Abs[z]	absolute value $|z|$
Arg[z]	the argument ϕ in $|z|e^{i\phi}$

Complex number operations.

1.4 Algebraic Calculations

+■ 1.4.6 Advanced Topic: Simplifying with Assumptions

+ Simplify[*expr*, *assum*]	simplify *expr* with assumptions

Simplifying with assumptions.

Mathematica does not automatically simplify this, since it is only true for some values of x.	*In[1]:=* **Simplify[Sqrt[x^2]]** *Out[1]=* $\sqrt{x^2}$
$\sqrt{x^2}$ is equal to x for $x \geq 0$, but not otherwise.	*In[2]:=* **{Sqrt[4^2], Sqrt[(-4)^2]}** *Out[2]=* {4, 4}
This tells Simplify to make the assumption x > 0, so that simplification can proceed.	*In[3]:=* **Simplify[Sqrt[x^2], x > 0]** *Out[3]=* x
No automatic simplification can be done on this expression.	*In[4]:=* **2 a + 2 Sqrt[a - Sqrt[-b]] Sqrt[a + Sqrt[-b]]** *Out[4]=* $2a + 2\sqrt{a - \sqrt{-b}} \ \sqrt{a + \sqrt{-b}}$
If *a* and *b* are assumed to be positive, the expression can however be simplified.	*In[5]:=* **Simplify[%, a > 0 && b > 0]** *Out[5]=* $2\left(a + \sqrt{a^2 + b}\right)$
Here is a simple example involving trigonometric functions.	*In[6]:=* **Simplify[ArcSin[Sin[x]], -Pi/2 < x < Pi/2]** *Out[6]=* x

Element[*x*, *dom*]	state that *x* is an element of the domain *dom*
Element[{x_1, x_2, ... }, *dom*]	state that all the x_i are elements of the domain *dom*
Reals	real numbers
Integers	integers
Primes	prime numbers

Some domains used in assumptions.

This simplifies $\sqrt{x^2}$ assuming that x is a real number.	*In[7]:=* **Simplify[Sqrt[x^2], Element[x, Reals]]** *Out[7]=* Abs[x]

This simplifies the sine assuming that *n* is an integer.

```
In[8]:= Simplify[Sin[x + 2 n Pi], Element[n, Integers]]
Out[8]= Sin[x]
```

With the assumptions given, Fermat's Little Theorem can be used.

```
In[9]:= Simplify[Mod[a^p, p], Element[a, Integers]
            && Element[p, Primes]]
Out[9]= Mod[a, p]
```

This uses the fact that sin(*x*), but not arcsin(*x*), is real when *x* is real.

```
In[10]:= Simplify[Re[{Sin[x], ArcSin[x]}], Element[x, Reals]]
Out[10]= {Sin[x], Re[ArcSin[x]]}
```

1.5 Symbolic Mathematics

⊹■ 1.5.11 Integral Transforms

⊹ LaplaceTransform[*expr*, *t*, *s*]	find the Laplace transform of *expr*
⊹ InverseLaplaceTransform[*expr*, *s*, *t*]	find the inverse Laplace transform of *expr*

Laplace transforms.

This computes a Laplace transform.

In[1]:= **LaplaceTransform[t^3 Exp[a t], t, s]**

$$Out[1]= \frac{6}{(-a+s)^4}$$

Here is the inverse transform.

In[2]:= **InverseLaplaceTransform[%, s, t]**

$$Out[2]= e^{a\,t}\, t^3$$

⊹ FourierTransform[*expr*, *t*, *w*]	find the symbolic Fourier transform of *expr*
⊹ InverseFourierTransform[*expr*, *w*, *t*]	find the inverse Fourier transform of *expr*

Fourier transforms.

This computes a Fourier transform.

In[3]:= **FourierTransform[t^4 Exp[-t^2], t, w]**

$$Out[3]= \frac{e^{-\frac{w^2}{4}}\,(12 - 12\,w^2 + w^4)}{16\,\sqrt{2}}$$

Here is the inverse transform.

In[4]:= **InverseFourierTransform[%, w, t]**

$$Out[4]= e^{-t^2}\, t^4$$

Note that in the scientific and technical literature many different conventions are used for defining Fourier transforms. Page 868 of the complete *Mathematica Book* describes the setup in *Mathematica*.

■ 1.5.12 Packages for Symbolic Mathematics

There are many *Mathematica* packages which implement symbolic mathematical operations. This section describes a few examples drawn from the standard set of packages distributed with *Mathematica*.

As discussed in Section 1.3.10 of the complete *Mathematica Book,* some copies of *Mathematica* may be set up so that the functions described here are automatically loaded into *Mathematica* if they are ever needed.

Solving Inequalities

`<<Algebra'InequalitySolve'`	load the inequality solving package
`InequalitySolve[`*ineq*`, `*x*`]`	solve inequalities with respect to *x*

Solving inequalities.

This loads the inequality solving package. In some versions of *Mathematica*, you may not need to load the package explicitly.

```
In[1]:= <<Algebra'InequalitySolve'
```

The solution to this inequality is a pair of intervals.

```
In[2]:= InequalitySolve[Abs[x-1](x^2-3) > 3, x]
```

$$Out[2]= \ x < -2 \ || \ x > \frac{1}{2}\left(1+\sqrt{13}\right)$$

Solving Recurrence Relations

`<<DiscreteMath'RSolve'`	load the recurrence relation solving package
`RSolve[`*eqn*`, `*a*`[`*n*`], `*n*`]`	solve the recurrence relation *eqn* for *a*[*n*]

Solving recurrence relations.

This loads the recurrence relation solving package. In some versions of *Mathematica*, you may not need to load the package explicitly.

```
In[1]:= <<DiscreteMath'RSolve'
```

This solves the recurrence relation for the factorial function.

```
In[2]:= RSolve[{a[n]==n a[n-1], a[1]==1}, a[n], n]
Out[2]= {{a[n] → n!}}
```
Gamma [1+n]

Here is the solution to a slightly more complicated recurrence relation.

```
In[3]:= RSolve[{a[n]==a[n-1] + 3a[n-2], a[0]==a[1]==1}, a[n], n]
```

$$Out[3]= \ \left\{\left\{a[n] \to -\frac{2^{-1-n}\left(\left(1-\sqrt{13}\right)^{1+n}-\left(1+\sqrt{13}\right)^{1+n}\right)}{\sqrt{13}}\right\}\right\}$$

1.8 Lists

■ 1.8.4 Getting Pieces of Lists

First[*list*]	the first element in *list*
Last[*list*]	the last element
Part[*list*, *n*] or *list*[[*n*]]	the nth element
Part[*list*, -*n*] or *list*[[-*n*]]	the nth element from the end
Part[*list*, {n_1, n_2, ... }] or *list*[[{n_1, n_2, ... }]]	the list of elements at positions n_1, n_2, ...

Picking out elements of lists.

We will use this list for the examples.

```
In[1]:= t = {a,b,c,d,e,f,g}
Out[1]= {a, b, c, d, e, f, g}
```

Here is the last element of t.

```
In[2]:= Last[t]
Out[2]= g
```

This gives the third element.

```
In[3]:= t[[3]]
Out[3]= c
```

This gives a list of the first and fourth elements.

```
In[4]:= t[[ {1, 4} ]]
Out[4]= {a, d}
```

Take[*list*, *n*]	the first *n* elements in *list*
Take[*list*, -*n*]	the last *n* elements
Take[*list*, {*m*, *n*}]	elements *m* through *n* (inclusive)
Take[*list*, {*m*, *n*, *s*}]	elements *m* through *n* in steps of *s*
Rest[*list*]	*list* with its first element dropped
Drop[*list*, *n*]	*list* with its first *n* elements dropped
Drop[*list*, -*n*]	*list* with its last *n* elements dropped
Drop[*list*, {*m*, *n*}]	*list* with elements *m* through *n* dropped
Drop[*list*, {*m*, *n*, *s*}]	*list* with elements *m* through *n* in steps of *s* dropped

Picking out sequences in lists.

This gives the first three elements of the list t defined above.	*In[5]:=* **Take[t, 3]**
	Out[5]= {a, b, c}
This gives the last three elements.	*In[6]:=* **Take[t, -3]**
	Out[6]= {e, f, g}
This gives elements 2 through 5 inclusive.	*In[7]:=* **Take[t, {2, 5}]**
	Out[7]= {b, c, d, e}
This gives elements 3 through 7 in steps of 2.	*In[8]:=* **Take[t, {3, 7, 2}]**
	Out[8]= {c, e, g}
This gives t with the first element dropped.	*In[9]:=* **Rest[t]**
	Out[9]= {b, c, d, e, f, g}
This gives t with its first three elements dropped.	*In[10]:=* **Drop[t, 3]**
	Out[10]= {d, e, f, g}
This gives t with only its third element dropped.	*In[11]:=* **Drop[t, {3, 3}]**
	Out[11]= {a, b, d, e, f, g}

Part[*list*, i, j, ...]   or   *list*[[i, j, ...]]
the element *list*[[i]][[j]] ...

Part[*list*, {i_1, i_2, ... }, {j_1, j_2, ... }, ...]   or   *list*[[{i_1, i_2, ... }, {j_1, j_2, ... }, ...]]
the list of elements obtained by picking out parts i_1, i_2, ...
at the first level, etc.

◦ Part[*list*, ... , All, ...]   or   *list*[[... , All, ...]]
a list of elements including all parts at a particular level

◦ Part[*list*, All, i]   or   *list*[[All, i]]
the ith "column" of *list*

Extracting parts of nested lists.

Here is a list of lists.	*In[12]:=* t = {{a, b, c}, {d, e, f}}
	Out[12]= {{a, b, c}, {d, e, f}}
This picks out the first sublist.	*In[13]:=* t[[1]]
	Out[13]= {a, b, c}
This picks out the second element in the first sublist.	*In[14]:=* t[[1, 2]]
	Out[14]= b
This is equivalent to t[[1, 2]], but is clumsier to write.	*In[15]:=* t[[1]][[2]]
	Out[15]= b
This gives a list containing two copies of the second part of t, followed by one copy of the first part.	*In[16]:=* t[[{2, 2, 1}]]
	Out[16]= {{d, e, f}, {d, e, f}, {a, b, c}}
For each of the parts picked out on the previous line, this gives a list of their second and third parts.	*In[17]:=* t[[{2, 2, 1}, {2, 3}]]
	Out[17]= {{e, f}, {e, f}, {b, c}}
This in effect gives the second column of t.	*In[18]:=* t[[All, 2]]
	Out[18]= {b, e}
This gives a list of the second and third columns.	*In[19]:=* t[[All, {2, 3}]]
	Out[19]= {{b, c}, {e, f}}

+ `Extract[`*list*`, {`*i*`, ` *j*`, ... }]`	the element at position {*i*, *j*, ... } in *list*
+ `Extract[`*list*`, {{`i_1`, `j_1`, ... }, {`i_2`, `j_2`, ... }, ... }]`	
	the list of elements at positions {i_1, j_1, ... }, {i_2, j_2, ... }, ...

Another way to extract parts of nested lists.

This extracts the element at position {2, 1} in t.	`In[20]:=` **`Extract[t, {2, 1}]`**
	`Out[20]=` d
This extracts a list of three elements from t.	`In[21]:=` **`Extract[t, {{1, 1}, {2, 2}, {2, 3}}]`**
	`Out[21]=` {a, e, f}

+ `Take[`*list*`, {`m_1`, `n_1`}, {`m_2`, `n_2`}, ... ]`	
	the nested list obtained by taking parts m_i through n_i of *list* at level *i*
+ `Drop[`*list*`, {`m_1`, `n_1`}, {`m_2`, `n_2`}, ... ]`	
	the nested list obtained by dropping parts m_i through n_i from *list* at level *i*

Picking out blocks in nested lists.

Here is a 3 × 3 matrix.	`In[22]:=` **`m = {{a, b, c}, {d, e, f}, {g, h, i}};`**
This extracts the bottom-right 2 × 2 submatrix.	`In[23]:=` **`Take[m, {2, 3}, {2, 3}]`**
	`Out[23]=` {{e, f}, {h, i}}
This leaves only the top-left 1 × 1 submatrix.	`In[24]:=` **`Drop[m, {2, 3}, {2, 3}]`**
	`Out[24]=` {{a}}
This gives the same result.	`In[25]:=` **`Drop[m, -2, -2]`**
	`Out[25]=` {{a}}

Section 2.1.5 of the complete *Mathematica Book* shows how all the functions in this section can be generalized to work not only on lists, but on any *Mathematica* expressions.

The functions in this section allow you to pick out pieces that occur at particular positions in lists. Section 2.3.2 of the complete *Mathematica Book* shows how you can use functions like `Select` and `Cases` to pick out elements of lists based not on their positions, but instead on their properties.

■ 1.8.6 Adding, Removing and Modifying List Elements

Prepend[*list*, *element*]	add *element* at the beginning of *list*
Append[*list*, *element*]	add *element* at the end of *list*
Insert[*list*, *element*, *i*]	insert *element* at position *i* in *list*
Insert[*list*, *element*, -*i*]	insert at position *i* counting from the end of *list*
Insert[*list*, *element*, {*i*, *j*, ... }]	insert at position *i*, *j*, ... in *list*
Delete[*list*, *i*]	delete the element at position *i* in *list*
Delete[*list*, {*i*, *j*, ... }]	delete at position *i*, *j*, ... in *list*

Functions for adding and deleting elements in lists.

This gives a list with x prepended.

```
In[1]:= Prepend[{a, b, c}, x]
Out[1]= {x, a, b, c}
```

This adds x at the end.

```
In[2]:= Append[{a, b, c}, x]
Out[2]= {a, b, c, x}
```

This inserts x so that it becomes element number 2.

```
In[3]:= Insert[{a, b, c}, x, 2]
Out[3]= {a, x, b, c}
```

Negative numbers count from the end of the list.

```
In[4]:= Insert[{a, b, c}, x, -2]
Out[4]= {a, b, x, c}
```

Delete removes an element from the list.

```
In[5]:= Delete[{a, b, c, d}, 3]
Out[5]= {a, b, d}
```

ReplacePart[*list*, *new*, *i*]	replace the element at position *i* in *list* with *new*
ReplacePart[*list*, *new*, -*i*]	replace at position *i* counting from the end
ReplacePart[*list*, *new*, {*i*, *j*, ... }]	replace *list*[[*i*, *j*, ...]] with *new*
ReplacePart[*list*, *new*, {{i_1, j_1, ... }, {i_2, ... }, ... }]	
	replace all parts *list*[[i_k, j_k, ...]] with *new*
ReplacePart[*list*, *new*, {{i_1, ... }, ... }, {n_1, n_2, ... }]	
	replace part *list*[[i_k, ...]] with *new*[[n_k]]

Modifying parts of lists.

This replaces the third element in the list with x.

```
In[6]:= ReplacePart[{a, b, c, d}, x, 3]
Out[6]= {a, b, x, d}
```

This replaces the first and fourth parts of the list. Notice the need for double lists in specifying multiple parts to replace.

```
In[7]:= ReplacePart[{a, b, c, d}, x, {{1}, {4}}]
Out[7]= {x, b, c, x}
```

Here is a 3 × 3 identity matrix.

```
In[8]:= IdentityMatrix[3]
Out[8]= {{1, 0, 0}, {0, 1, 0}, {0, 0, 1}}
```

This replaces the (2, 2) component of the matrix by x.

```
In[9]:= ReplacePart[%, x, {2, 2}]
Out[9]= {{1, 0, 0}, {0, x, 0}, {0, 0, 1}}
```

+	PadLeft[*list*, *length*]	make a list of the specified length by padding with zeros on the left
+	PadLeft[*list*, *length*, *x*]	pad by repeating the element *x*
+	PadLeft[*list*, *length*, {x_1, x_2, ... }]	pad by cyclically repeating the elements x_i
+	PadLeft[*list*, *length*, *list*]	pad by cyclically repeating the original list
+	PadLeft[*list*, *length*, *padding*, *m*]	leave a margin of *m* elements of padding on the right
+	PadRight[*list*, *length*, ...]	insert padding on the right

Padding lists.

This pads with zeros on the left to make a list of length 10.

```
In[10]:= PadLeft[{a, b, c}, 10]
Out[10]= {0, 0, 0, 0, 0, 0, 0, a, b, c}
```

This pads with copies of the sequence x, y.

```
In[11]:= PadLeft[{a, b, c}, 10, {x, y}]
Out[11]= {x, y, x, y, x, y, x, a, b, c}
```

This leaves a margin of 3 elements on the right.

```
In[12]:= PadLeft[{a, b, c}, 10, {x, y}, 3]
Out[12]= {y, x, y, x, a, b, c, x, y, x}
```

■ 1.8.11 Advanced Topic: Alignment and Padding in the Partitioning of Lists

What Partition does is to pick out a sequence of sublists from a list. A subtle but important issue is where to begin the first sublist and where to end the last sublist relative to the original list. The default is simply to drop any sublists that would "overhang" on either side. In general *Mathematica* allows you to specify the position in the first sublist that should correspond to the first element of the original list, and the first possible position in the last sublist that should be allowed to correspond to the last element of the original list. When this specification implies that there should be overhangs,

Mathematica by default fills in the necessary additional elements by treating the original list as cyclic. It also allows you instead to explicitly give a sequence of elements with which to "pad out" the list.

⊹ Partition[*list*, *n*, *d*] or Partition[*list*, *n*, *d*, {1, -1}]

   partition *list* keeping only sublists that do not overhang on either side

⊹ Partition[*list*, *n*, *d*, {1, 1}]   partition allowing an overhang at the end, treating *list* as cyclic

⊹ Partition[*list*, *n*, *d*, {-1, -1}]

   partition allowing an overhang at the beginning, treating *list* as cyclic

⊹ Partition[*list*, *n*, *d*, {-1, 1}]   partition allowing overhangs at both the beginning and end

⊹ Partition[*list*, *n*, *d*, {k_L, k_R}]   specify explicitly the alignments of the first and last sublists

⊹ Partition[*list*, *n*, *d*, {k_L, k_R}, *x*]

   pad when necessary by repeating the element *x*

⊹ Partition[*list*, *n*, *d*, {k_L, k_R}, {x_1, x_2, ... }]

   pad by cyclically repeating the elements x_i

⊹ Partition[*list*, *n*, *d*, {k_L, k_R}, {}]

   use no padding, making sublists shorter when elements are not available

Alignment and padding of sublists.

Here is a list.	In[1]:= **t = {a, b, c, d, e, f, g}** Out[1]= {a, b, c, d, e, f, g}
This partitions the list, stopping before any overhang occurs.	In[2]:= **Partition[t, 3, 1]** Out[2]= {{a, b, c}, {b, c, d}, {c, d, e}, {d, e, f}, {e, f, g}}
The last two sublists involve overhangs. The additional elements needed are filled in by assuming that the original list is cyclic.	In[3]:= **Partition[t, 3, 1, {1, 1}]** Out[3]= {{a, b, c}, {b, c, d}, {c, d, e}, {d, e, f}, {e, f, g}, {f, g, a}, {g, a, b}}
This fills in additional elements by padding with x.	In[4]:= **Partition[t, 3, 1, {1, 1}, x]** Out[4]= {{a, b, c}, {b, c, d}, {c, d, e}, {d, e, f}, {e, f, g}, {f, g, x}, {g, x, x}}
This pads by cyclically repeating the block x, y.	In[5]:= **Partition[t, 3, 1, {1, 1}, {x, y}]** Out[5]= {{a, b, c}, {b, c, d}, {c, d, e}, {d, e, f}, {e, f, g}, {f, g, y}, {g, y, x}}

This has overhangs at both the beginning and end.

```
In[6]:= Partition[t, 3, 1, {-1, 1}, x]
Out[6]= {{x, x, a}, {x, a, b}, {a, b, c}, {b, c, d}, {c, d, e},
         {d, e, f}, {e, f, g}, {f, g, x}, {g, x, x}}
```

This uses no padding, and yields shorter lists at the beginning and end.

```
In[7]:= Partition[t, 3, 1, {-1, 1}, {}]
Out[7]= {{a}, {a, b}, {a, b, c}, {b, c, d}, {c, d, e},
         {d, e, f}, {e, f, g}, {f, g}, {g}}
```

■ 1.8.13 Advanced Topic: Rearranging Nested Lists

You will encounter nested lists if you use matrices or generate multidimensional arrays and tables. Rearranging nested lists can be a complicated affair, and you will often have to experiment to get the right combination of commands.

Transpose[*list*]	interchange the top two levels of lists
Transpose[*list*, {m, n, ... }]	put first level at level m, second level at level n, ...
Flatten[*list*]	flatten out all levels in *list*
Flatten[*list*, n]	flatten out the top n levels in *list*
FlattenAt[*list*, i]	flatten out a sublist that appears at position i
FlattenAt[*list*, {i, j, ... }]	flatten out a sublist at position i, j, ...
FlattenAt[*list*, {{i_1, j_1, ... }, {i_2, ... }, ... }]	flatten out several sublists
RotateLeft[*list*, {n_1, n_2, ... }], RotateRight[*list*, {n_1, n_2, ... }]	rotate successive levels by n_i places
Partition[*list*, {n_1, n_2, ... }]	partition into blocks of size $n_1 \times n_2 \times ...$
PadLeft[*list*, {s_1, s_2, ... }], PadRight[*list*, {s_1, s_2, ... }]	pad level i to be length s_i

Functions for rearranging nested lists.

Here is a 3×2 array.

```
In[1]:= t = {{a, b}, {c, d}, {e, f}}
Out[1]= {{a, b}, {c, d}, {e, f}}
```

You can rearrange it to get a 2×3 array.

```
In[2]:= Transpose[t]
Out[2]= {{a, c, e}, {b, d, f}}
```

This "flattens out" sublists. You can think of it as effectively just removing the inner sets of braces.

```
In[3]:= Flatten[t]
Out[3]= {a, b, c, d, e, f}
```

Here is a 2 × 2 × 2 array.

```
In[4]:= t = Table[i^2 +j^2 +k^2, {i, 2}, {j, 2}, {k, 2}]
Out[4]= {{{3, 6}, {6, 9}}, {{6, 9}, {9, 12}}}
```

This flattens out all the levels.

```
In[5]:= Flatten[t]
Out[5]= {3, 6, 6, 9, 6, 9, 9, 12}
```

This flattens only the first level of sublists.

```
In[6]:= Flatten[t, 1]
Out[6]= {{3, 6}, {6, 9}, {6, 9}, {9, 12}}
```

This flattens out only the sublist that appears at position 2.

```
In[7]:= FlattenAt[{{a, b}, {c, d}}, 2]
Out[7]= {{a, b}, c, d}
```

There are many other operations you can perform on nested lists. We will discuss some more of them when we look at Map, Apply, Scan and Level in Part 2.

1.10 Input and Output in Notebooks

■ 1.10.4 Entering Formulas

character	short form	long form	symbol
π	⋮p⋮	\[Pi]	Pi
∞	⋮inf⋮	\[Infinity]	Infinity
°	⋮deg⋮	\[Degree]	Degree

Special forms for some common symbols. ⋮ stands for the key [ESC].

This is equivalent to Sin[60 Degree].

$In[1]:=$ **Sin[60°]**

$Out[1]=$ $\dfrac{\sqrt{3}}{2}$

Here is the long form of the input.

$In[2]:=$ **Sin[60 \[Degree]]**

$Out[2]=$ $\dfrac{\sqrt{3}}{2}$

You can enter the same input like this.

$In[3]:=$ **Sin[60 ⋮deg⋮]**

$Out[3]=$ $\dfrac{\sqrt{3}}{2}$

Here the angle is in radians.

$In[4]:=$ **Sin$\left[\dfrac{\pi}{3}\right]$**

$Out[4]=$ $\dfrac{\sqrt{3}}{2}$

special characters	short form	long form	ordinary characters
$x \leq y$	x :<=: y	x \[LessEqual] y	x <= y
$x \geq y$	x :>=: y	x \[GreaterEqual] y	x >= y
$x \neq y$	x :!=: y	x \[NotEqual] y	x != y
✦ $x \in y$	x :elem: y	x \[Element] y	Element[x, y]
$x \rightarrow y$	x :->: y	x \[Rule] y	x -> y

Special forms for a few operators. Pages 74–79 give a complete list.

Here the replacement rule is entered using two ordinary characters, as ->.

In[5]:= x/(x+1) /. x -> 3 + y

Out[5]= $\dfrac{3 + y}{4 + y}$

This means exactly the same.

In[6]:= x/(x+1) /. x \[Rule] 3 + y

Out[6]= $\dfrac{3 + y}{4 + y}$

As does this.

In[7]:= x/(x+1) /. x → 3 + y

Out[7]= $\dfrac{3 + y}{4 + y}$

Or this.

In[8]:= x/(x+1) /. x :->: 3 + y

Out[8]= $\dfrac{3 + y}{4 + y}$

The special arrow form → is by default also used for output.

In[9]:= Solve[x^2 == 1, x]

Out[9]= {{x → -1}, {x → 1}}

special characters	short form	long form	ordinary characters
$x \div y$	x :div: y	x \[Divide] y	x / y
$x \times y$	x :*: y	x \[Times] y	x * y
$x \times y$	x :cross: y	x \[Cross] y	Cross[x, y]
$x == y$	x :==: y	x \[Equal] y	$x == y$
$x \wedge y$	x :&&: y	x \[And] y	x && y
$x \vee y$	x :||: y	x \[Or] y	x || y
$\neg\, x$	:!: x	\[Not] x	!x
$x \Rightarrow y$	x :=>: y	x \[Implies] y	Implies[x, y]
$x \cup y$	x :un: y	x \[Union] y	Union[x, y]
$x \cap y$	x :inter: y	x \[Intersection] y	Intersection[x, y]
xy	x :,: y	x \[InvisibleComma] y	x,y
fx	f :@: x	f \[InvisibleApplication] x	f@x or $f[x]$

Some operators with special forms used for input but not output.

Mathematica understands $\div$, but does not use it by default for output.

$In[10]:=$ **x** $\div$ **y**

$Out[10]=$ $\dfrac{\text{x}}{\text{y}}$

The forms of input discussed so far in this section use special characters, but otherwise just consist of ordinary one-dimensional lines of text. *Mathematica* notebooks, however, also make it possible to use two-dimensional forms of input.

two-dimensional	one-dimensional	
x^y	x ^ y	power
$\dfrac{x}{y}$	x / y	division
$\sqrt{x}$	Sqrt[x]	square root
$\sqrt[n]{x}$	x ^ (1/n)	n^{th} root
$\displaystyle\sum_{i=imin}^{imax} f$	Sum[f, {i, imin, imax}]	sum
$\displaystyle\prod_{i=imin}^{imax} f$	Product[f, {i, imin, imax}]	product
$\displaystyle\int f\,dx$	Integrate[f, x]	indefinite integral
$\displaystyle\int_{xmin}^{xmax} f\,dx$	Integrate[f, {x, xmin, xmax}]	definite integral
$\partial_x f$	D[f, x]	partial derivative
$\partial_{x,y} f$	D[f, x, y]	multivariate partial derivative
$expr_{[\![i,j,\dots]\!]}$	Part[expr, i, j, ...]	part extraction

Some two-dimensional forms that can be used in *Mathematica* notebooks.

You can enter two-dimensional forms using any of the mechanisms discussed on pages 175–179 of the complete *Mathematica Book*. Note that upper and lower limits for sums and products must be entered as overscripts and underscripts—not superscripts and subscripts.

This enters an indefinite integral. Note the use of ⦂dd⦂ to enter the "differential d".	$In[11]:= $ ⦂int⦂ f[x] ⦂dd⦂ x
	$Out[11]= \displaystyle\int f[x]\,dx$
Here is an indefinite integral that can be explicitly evaluated.	$In[12]:= \displaystyle\int$ Exp[-x²] dx
	$Out[12]= \dfrac{1}{2}\sqrt{\pi}\ \text{Erf}[x]$

Here is the usual *Mathematica* input for this integral.	*In[13]:=* `Integrate[Exp[-x^2], x]`
	Out[13]= $\frac{1}{2} \sqrt{\pi}$ `Erf[x]`
This enters exactly the same integral.	*In[14]:=* `\!\( \[Integral] Exp[-x\^2] \[DifferentialD]x \)`
	Out[14]= $\frac{1}{2} \sqrt{\pi}$ `Erf[x]`

short form	long form	
⋮sum⋮	`\[Sum]`	summation sign $\sum$
⋮prod⋮	`\[Product]`	product sign $\prod$
⋮int⋮	`\[Integral]`	integral sign $\int$
⋮dd⋮	`\[DifferentialD]`	special $\text{d}\!\!\!\text{l}$ for use in integrals
⋮pd⋮	`\[PartialD]`	partial derivative operator ∂
⋮[[⋮, ⋮]]⋮	`\[LeftDoubleBracket], \[RightDoubleBracket]` part brackets

Some special characters used in entering formulas. Section 3.10 of the complete *Mathematica Book* gives a complete list.

You should realize that even though a summation sign can look almost identical to a capital sigma it is treated in a very different way by *Mathematica*. The point is that a sigma is just a letter; but a summation sign is an operator which tells *Mathematica* to perform a Sum operation.

Capital sigma is just a letter.	*In[15]:=* `a + \[CapitalSigma]^2`
	Out[15]= $a + \Sigma^2$
A summation sign, on the other hand, is an operator.	*In[16]:=* ESC **sum** ESC CTRL + n=0 CTRL % m CTRL _ `1/f[n]`
	Out[16]= $\displaystyle\sum_{n=0}^{m} \frac{1}{f[n]}$
Here is another way to enter the same input.	*In[17]:=* `\!\( \[Sum] \+ \( n = 0 \) \% m 1 \/ f[n] \)`
	Out[17]= $\displaystyle\sum_{n=0}^{m} \frac{1}{f[n]}$

Much as *Mathematica* distinguishes between a summation sign and a capital sigma, it also distinguishes between an ordinary d and the special "differential d" $\text{d}\!\!\!\text{l}$ that is used in the standard notation for integrals. It is crucial that you use this differential $\text{d}\!\!\!\text{l}$—entered as ESC dd ESC—when you type in an integral. If you try to use an ordinary d, *Mathematica* will just interpret this as a symbol called d—it will not understand that you are entering the second part of an integration operator.

This computes the derivative of x^n.

In[18]:= ∂_x **x^n**

Out[18]= $n\,x^{-1+n}$

Here is the same derivative specified in ordinary one-dimensional form.

In[19]:= **D[x^n, x]**

Out[19]= $n\,x^{-1+n}$

This computes the third derivative.

In[20]:= $\partial_{x,x,x}$ **x^n**

Out[20]= $(-2+n)\,(-1+n)\,n\,x^{-3+n}$

Here is the equivalent one-dimensional input form.

In[21]:= **D[x^n, x, x, x]**

Out[21]= $(-2+n)\,(-1+n)\,n\,x^{-3+n}$

■ 1.10.10 Mixing Text and Formulas

The simplest way to mix text and formulas in a *Mathematica* notebook is to put each kind of material in a separate cell. Sometimes, however, you may want to embed a formula within a cell of text, or vice versa.

CTRL (or CTRL 9	begin entering a formula within text, or text within a formula
CTRL) or CTRL 0	end entering a formula within text, or text within a formula

Entering a formula within text, or vice versa.

Here is a notebook with formulas embedded in a text cell.

This is a text cell, but it can contain formulas such as $\int \frac{1}{x^3-1}\,dx$ or $-\frac{\log(x^2+x+1)}{6} - \frac{\tan^{-1}\left(\frac{2x+1}{\sqrt{3}}\right)}{\sqrt{3}} + \frac{\log(x-1)}{3}$. The formulas flow with the text.

Mathematica notebooks often contain both formulas that are intended for actual evaluation by *Mathematica*, and ones that are intended just to be read in a more passive way.

When you insert a formula in text, you can use the Convert to StandardForm and Convert to TraditionalForm menu items within the formula to convert it to `StandardForm` or `TraditionalForm`. `StandardForm` is normally appropriate whenever the formula is thought of as a *Mathematica* program fragment.

In general, however, you can use exactly the same mechanisms for entering formulas, whether or not they will ultimately be given as *Mathematica* input.

You should realize, however, that to make the detailed typography of typical formulas look as good as possible, *Mathematica* automatically does things such as inserting spaces around certain operators. But these kinds of adjustments can potentially be inappropriate if you use notation in very different ways from the ones *Mathematica* is expecting.

In such cases, you may have to make detailed typographical adjustments by hand, using the mechanisms discussed on page 428 of the complete *Mathematica Book*.

1.11 Files and External Operations

■ 1.11.3 Importing and Exporting Data

Import["*file*", "Table"]	import a table of data from a file
Export["*file*", *list*, "Table"]	export *list* to a file as a table of data

Importing and exporting tabular data.

This exports an array of numbers to the file out.dat.	In[1]:= **Export["out.dat", {{5.7, 4.3}, {-1.2, 7.8}}]**
	Out[1]= /mnt/Pubs7/TeX3.1/Addendum/English/out.dat
Here are the contents of the file out.dat.	In[2]:= **!!out.dat**
	5.7 4.3
	-1.2 7.8
This imports the contents of out.dat as a table of data.	In[3]:= **Import["out.dat", "Table"]**
	Out[3]= {{5.7, 4.3}, {-1.2, 7.8}}

Import["*file*", "Table"] will handle many kinds of tabular data, automatically deducing the details of the format whenever possible. Export["*file*", *list*, "Table"] writes out data separated by spaces, with numbers given in C or Fortran-like form, as in 2.3E5 and so on.

Import["*name.ext*"]	import data assuming a format deduced from the file name
Export["*name.ext*", *expr*]	export data in a format deduced from the file name

Importing and exporting general data.

Import and Export can handle not only tabular data, but also data corresponding to graphics, sounds and even whole documents. Import and Export can often deduce the appropriate format for data simply by looking at the extension of the file name for the file in which the data is being stored. Sections 2.9.19 and 2.11.7 discuss in more detail how Import and Export work. Note that you can also use Import and Export to manipulate raw files of binary data.

This imports a graphic in JPEG format.	In[4]:= **Import["turtle.jpg"]**
	Out[4]= ‐Graphics‐

This displays the graphic.   *In[5]:=* **Show[%]**

+	$ImportFormats	import formats supported on your system
+	$ExportFormats	export formats supported on your system

Finding the complete list of supported import and export formats.

■ 1.11.4 Exporting Graphics and Sounds

Mathematica allows you to export graphics and sounds in a wide variety of formats. If you use the notebook front end for *Mathematica*, then you can typically just copy and paste graphics and sounds directly into other programs using the standard mechanism available on your computer system.

+ Export["*name.ext*", *graphics*]	export graphics to a file in a format deduced from the file name
+ Export["*file*", *graphics*, "*format*"]	export graphics in the specified format
+ Export["!*command*", *graphics*, "*format*"]	export graphics to an external command

Exporting *Mathematica* graphics and sounds.

graphics formats	"EPS", "TIFF", "GIF", "JPEG", "PDF", "AI", etc.
sound formats	"SND", "WAV", "AIFF", "AU", etc.

Some common formats for graphics and sounds. Pages 30 and 31 give a complete list.

This generates a plot.   *In[1]:=* **Plot[Sin[x] + Sin[Sqrt[2] x], {x, 0, 10}]**

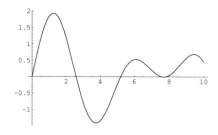

This exports the plot to a file in   *In[2]:=* **Export["sinplot.eps", %]**
Encapsulated PostScript format.
                                            Out[2]= /mnt/Pubs7/TeX3.1/Addendum/English/sinplot.eps

■ 1.11.5 Exporting Formulas from Notebooks

Here is a cell containing a formula.

$$-\frac{ArcTan[\frac{1+2x}{\sqrt{3}}]}{\sqrt{3}} + \frac{Log[-1+x]}{3} - \frac{Log[1+x+x^2]}{6}$$

This is what you get if you copy the   `\!\(-\(ArcTan[\(1 + 2 x\)\/\@3]\/\@3\) + Log[-1 + x]\/3`
formula and paste it into an external   `- Log[1 + x + x\^2]\/6\)`
text-based program.

Pasting the text back into a notebook
immediately reproduces the original $$-\frac{ArcTan[\frac{1+2x}{\sqrt{3}}]}{\sqrt{3}} + \frac{Log[-1+x]}{3} - \frac{Log[1+x+x^2]}{6}$$
formula.

Mathematica allows you to export formulas both textually and visually. You can use `Export` to tell
Mathematica to write a visual representation of a formula into a file.

⊹ `Export["`*file*`.eps", ToBoxes[`*expr*`]]`
                            save the visual form of *expr* to a file in EPS format

⊹ `Export["`*file*`", ToBoxes[`*expr*`], "`*format*`"]`
                            save the visual form of *expr* in the specified format

---

Exporting expressions in visual form.

2.2 Functional Operations

■ 2.2.2 Applying Functions Repeatedly

Many programs you write will involve operations that need to be iterated several times. Nest and NestList are powerful constructs for doing this.

Nest[f, x, n]	apply the function f nested n times to x
NestList[f, x, n]	generate the list {x, $f[x]$, $f[f[x]]$, ... }, where f is nested up to n deep

Applying functions of one argument repeatedly.

Nest[f, x, n] takes the "name" f of a function, and applies the function n times to x.

```
In[1]:= Nest[f, x, 4]
Out[1]= f[f[f[f[x]]]]
```

This makes a list of each successive nesting.

```
In[2]:= NestList[f, x, 4]
Out[2]= {x, f[x], f[f[x]], f[f[f[x]]], f[f[f[f[x]]]]}
```

Here is a simple function.

```
In[3]:= recip[x_] := 1/(1 + x)
```

You can iterate the function using Nest.

```
In[4]:= Nest[recip, x, 3]
```

$$Out[4]= \cfrac{1}{1 + \cfrac{1}{1 + \cfrac{1}{1+x}}}$$

Nest and NestList allow you to apply functions a fixed number of times. Often you may want to apply functions until the result no longer changes. You can do this using FixedPoint and FixedPointList.

FixedPoint[f, x]	apply the function f repeatedly until the result no longer changes
FixedPointList[f, x]	generate the list {x, $f[x]$, $f[f[x]]$, ... }, stopping when the elements no longer change

Applying functions until the result no longer changes.

Here is a function that takes one step in Newton's approximation to $\sqrt{3}$.

```
In[5]:= newton3[x_] := N[ 1/2 ( x + 3/x ) ]
```

Here are five successive iterates of the function, starting at $x = 1$.

```
In[6]:= NestList[newton3, 1.0, 5]
Out[6]= {1., 2., 1.75, 1.73214, 1.73205, 1.73205}
```

Using the function FixedPoint, you can automatically continue applying newton3 until the result no longer changes.

```
In[7]:= FixedPoint[newton3, 1.0]
Out[7]= 1.73205
```

Here is the sequence of results.

```
In[8]:= FixedPointList[newton3, 1.0]
Out[8]= {1., 2., 1.75, 1.73214, 1.73205, 1.73205, 1.73205}
```

∗	NestWhile[f, x, *test*]	apply the function f repeatedly until applying *test* to the result no longer yields True
∗	NestWhileList[f, x, *test*]	generate the list {x, $f[x]$, $f[f[x]]$, ... }, stopping when applying *test* to the result no longer yields True
∗	NestWhile[f, x, *test*, m], NestWhileList[f, x, *test*, m]	supply the m most recent results as arguments for *test* at each step
∗	NestWhile[f, x, *test*, All], NestWhileList[f, x, *test*, All]	supply all results so far as arguments for *test*

Applying functions repeatedly until a test fails.

Here is a function which divides a number by 2.

```
In[9]:= divide2[n_] := n/2
```

This repeatedly applies divide2 until the result is no longer an even number.

```
In[10]:= NestWhileList[divide2, 123456, EvenQ]
Out[10]= {123456, 61728, 30864, 15432, 7716, 3858, 1929}
```

This repeatedly applies newton3, stopping when two successive results are no longer considered unequal, just as in FixedPointList.

```
In[11]:= NestWhileList[newton3, 1.0, Unequal, 2]
Out[11]= {1., 2., 1.75, 1.73214, 1.73205, 1.73205, 1.73205}
```

This goes on until the first time a result that has been seen before reappears.

```
In[12]:= NestWhileList[Mod[5 #, 7]&, 1, Unequal, All]
Out[12]= {1, 5, 4, 6, 2, 3, 1}
```

Operations such as Nest take a function f of one argument, and apply it repeatedly. At each step, they use the result of the previous step as the new argument of f.

It is important to generalize this notion to functions of two arguments. You can again apply the function repeatedly, but now each result you get supplies only one of the new arguments you need. A convenient approach is to get the other argument at each step from the successive elements of a list.

FoldList[*f*, *x*, {*a*, *b*, ... }]	create the list {*x*, *f*[*x*, *a*], *f*[*f*[*x*, *a*], *b*], ... }
Fold[*f*, *x*, {*a*, *b*, ... }]	give the last element of the list produced by FoldList[*f*, *x*, {*a*, *b*, ... }]

Ways to repeatedly apply functions of two arguments.

Here is an example of what FoldList does.

```
In[13]:= FoldList[f, x, {a, b, c}]

Out[13]= {x, f[x, a], f[f[x, a], b], f[f[f[x, a], b], c]}
```

Fold gives the last element of the list produced by FoldList.

```
In[14]:= Fold[f, x, {a, b, c}]

Out[14]= f[f[f[x, a], b], c]
```

This gives a list of cumulative sums.

```
In[15]:= FoldList[Plus, 0, {a, b, c}]

Out[15]= {0, a, a + b, a + b + c}
```

Using Fold and FoldList you can write many elegant and efficient programs in *Mathematica*. In some cases, you may find it helpful to think of Fold and FoldList as producing a simple nesting of a family of functions indexed by their first argument.

This defines a function nextdigit.

```
In[16]:= nextdigit[a_, b_] := 10 a + b
```

Here is a rather elegant definition of a function that gives the number corresponding to a list of digits in base 10.

```
In[17]:= tonumber[digits_] := Fold[nextdigit, 0, digits]
```

Here is an example of tonumber in action.

```
In[18]:= tonumber[{1, 3, 7, 2, 9, 1}]

Out[18]= 137291
```

■ 2.2.3 Applying Functions to Lists and Other Expressions

In an expression like f[{a, b, c}] you are giving a list as the argument to a function. Often you need instead to apply a function directly to the elements of a list, rather than to the list as a whole. You can do this in *Mathematica* using Apply.

This makes each element of the list an argument of the function f.

```
In[1]:= Apply[f, {a, b, c}]

Out[1]= f[a, b, c]
```

This gives Plus[a, b, c] which yields the sum of the elements in the list.

```
In[2]:= Apply[Plus, {a, b, c}]

Out[2]= a + b + c
```

Here is the definition of the statistical mean, written using Apply.

```
In[3]:= mean[list_] := Apply[Plus, list] / Length[list]
```

Apply[*f*, {*a*, *b*, ... }]	apply *f* to a list, giving *f*[*a*, *b*, ...]
Apply[*f*, *expr*] or *f* @@ *expr*	apply *f* to the top level of an expression
+ Apply[*f*, *expr*, {1}] or *f* @@@ *expr*	
	apply *f* at the first level in an expression
Apply[*f*, *expr*, *lev*]	apply *f* at the specified levels in an expression

Applying functions to lists and other expressions.

What Apply does in general is to replace the head of an expression with the function you specify. Here it replaces Plus by List.

```
In[4]:= Apply[List, a + b + c]
Out[4]= {a, b, c}
```

Here is a matrix.

```
In[5]:= m = {{a, b, c}, {b, c, d}}
Out[5]= {{a, b, c}, {b, c, d}}
```

Using Apply without an explicit level specification replaces the top-level list with f.

```
In[6]:= Apply[f, m]
Out[6]= f[{a, b, c}, {b, c, d}]
```

This applies f only to parts of m at level 1.

```
In[7]:= Apply[f, m, {1}]
Out[7]= {f[a, b, c], f[b, c, d]}
```

This applies f at levels 0 through 1.

```
In[8]:= Apply[f, m, {0, 1}]
Out[8]= f[f[a, b, c], f[b, c, d]]
```

2.9 The Structure of Graphics and Sound

■ 2.9.19 Exporting Graphics and Sounds

Export["*name.ext*", *graphics*]	export graphics in a format deduced from the file name
Export["*file*", *graphics*, "*format*"]	export graphics in the specified format
Export["*file*", {g_1, g_2, ... }, ...]	export a sequence of graphics for an animation
ExportString[*graphics*, "*format*"]	generate a string representation of exported graphics

Exporting graphics and sounds.

"MPS"	*Mathematica* abbreviated PostScript (.mps)
"EPS"	Encapsulated PostScript (.eps)
"PDF"	Adobe Acrobat portable document format (.pdf)
"AI"	Adobe Illustrator format (.ai)
"PCL"	Hewlett-Packard printer control language (.pcl)
"PICT"	Macintosh PICT
"WMF"	Microsoft Windows metafile format (.wmf)
"TIFF"	TIFF (.tif, .tiff)
"GIF"	GIF and animated GIF (.gif)
"JPEG"	JPEG (.jpg, .jpeg)
"PNG"	PNG format (.png)
"BMP"	Microsoft bitmap format (.bmp)
"EPSI"	Encapsulated PostScript with image preview (.epsi)
"EPSTIFF"	Encapsulated PostScript with TIFF preview
"PSImage"	PostScript image format (.psi)
"XBitmap"	X window system bitmap (.xbm)
"PBM"	portable bitmap format (.pbm)
"PPM"	portable pixmap format (.ppm)
"PGM"	portable graymap format (.pgm)
"PNM"	portable anymap format (.pnm)
"MGF"	*Mathematica* system-independent raster graphics format

Typical graphics formats supported by *Mathematica*. The first group are resolution independent.

When you export a graphic outside of *Mathematica*, you usually have to specify the absolute size at which the graphic should be rendered. You can do this using the `ImageSize` option to `Export`.

`ImageSize->`x makes the width of the graphic be x printer's points; `ImageSize->72` xi thus makes the width xi inches. The default is to produce an image that is four inches wide. `ImageSize->`$\{x, y\}$ scales the graphic so that it fits in an $x \times y$ region.

+ ImageSize	Automatic	absolute image size in printer's points
+ ImageRotated	False	whether to rotate the image (landscape mode)
+ ImageResolution	Automatic	resolution in dpi for the image

Options for `Export`.

Within *Mathematica* graphics are manipulated in a way that is completely independent of the resolution of the computer screen or other output device on which the graphics will eventually be rendered.

Many programs and devices accept graphics in resolution-independent formats such as Encapsulated PostScript (EPS). But some require that the graphics be converted to rasters or bitmaps with a specific resolution. The `ImageResolution` option for `Export` allows you to determine what resolution in dots per inch (dpi) should be used. The lower you set this resolution, the lower the quality of the image you will get, but also the less memory the image will take to store. For screen display, typical resolutions are 72 dpi and above; for printers, 300 dpi and above.

"WAV"	Microsoft wave format (`.wav`)
"AU"	μ law encoding (`.au`)
"SND"	sound file format (`.snd`)
"AIFF"	AIFF format (`.aif`, `.aiff`)

Typical sound formats supported by *Mathematica*.

+ ■ 2.9.20 Importing Graphics and Sounds

Mathematica allows you not only to export graphics and sounds, but also to import them. With `Import` you can read graphics and sounds in a wide variety of formats, and bring them into *Mathematica* as *Mathematica* expressions.

⊹	Import["*name.ext*"]	import graphics from the file *name.ext* in a format deduced from the file name
⊹	Import["*file*", "*format*"]	import graphics in the specified format
⊹	ImportString["*string*", "*format*"]	import graphics from a string

Importing graphics and sounds.

This imports an image stored in JPEG format.

In[1]:= g = Import["ocelot.jpg"]

Out[1]= ·Graphics·

Here is the image.

In[2]:= Show[g]

This shows an array of four copies of the image.

In[3]:= Show[GraphicsArray[{{g, g}, {g, g}}]]

Import yields expressions with different structures depending on the type of data it reads. Typically you will need to know the structure if you want to manipulate the data that is returned.

Graphics[*primitives*, *opts*]	resolution-independent graphics
Graphics[Raster[*data*], *opts*]	resolution-dependent bitmap images
{*graphics*$_1$, *graphics*$_2$, ... }	animated graphics
Sound[SampledSoundList[*data*, *r*]]	
	sounds

Structures of expressions returned by Import.

This shows the overall structure of the graphics object imported above.

```
In[4]:= Short[InputForm[g]]

Out[4]//Short= Graphics[<<2>>]
```

This extracts the array of pixel values used.

```
In[5]:= d = g[[1, 1]] ;
```

Here are the dimensions of the array.

```
In[6]:= Dimensions[d]

Out[6]= {200, 200}
```

This shows the distribution of pixel values.

```
In[7]:= ListPlot[Sort[Flatten[d]]]
```

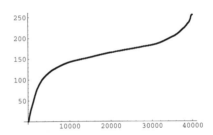

This shows a transformed version of the image.

```
In[8]:= Show[Graphics[Raster[d^2 / Max[d^2]]],
            AspectRatio->Automatic]
```

2.10 Manipulating Notebooks

■ 2.10.11 Advanced Topic: Options for Expression Input and Output

option	typical default value	
AutoIndent	Automatic	whether to indent after an explicit RETURN character is entered
DelimiterFlashTime	0.3	the time in seconds to flash a delimiter when a matching one is entered
ShowAutoStyles	True	whether to show automatic annotations of syntactic and other constructs
ShowCursorTracker	True	whether an elliptical spot should appear momentarily to guide the eye if the cursor position jumps
ShowSpecialCharacters	True	whether to replace \[*Name*] by a special character as soon as the] is entered
ShowStringCharacters	False	whether to display " when a string is entered
SingleLetterItalics	False	whether to put single-letter symbol names in italics
ZeroWidthTimes	False	whether to represent multiplication by a zero width character
InputAliases	{}	additional ꞉*name*꞉ aliases to allow
InputAutoReplacements	{"->"->"→", ... }	strings to automatically replace on input
AutoItalicWords	{"Mathematica", ... }	words to automatically put in italics
LanguageCategory	Automatic	what category of language to assume a cell contains for spell checking and hyphenation

Options associated with the interactive entering of expressions.

The options SingleLetterItalics and ZeroWidthTimes are typically set whenever a cell uses TraditionalForm.

Here is an expression entered with default options for a `StandardForm` input cell.

$$x^6 + 6 x^5 y + 15 x^4 y^2 + 20 x^3 y^3 + 15 x^2 y^4 + 6 x y^5 + y^6$$

Here is the same expression entered in a cell with `SingleLetterItalics->True` and `ZeroWidthTimes->True`.

$$x^6 + 6x^5y + 15x^4y^2 + 20x^3y^3 + 15x^2y^4 + 6xy^5 + y^6$$

Built into *Mathematica* are a large number of aliases for common special characters. `InputAliases` allows you to add your own aliases for further special characters or for any other kind of *Mathematica* input. A rule of the form `"name"->`*expr* specifies that ⫶*name*⫶ should immediately be replaced on input by *expr*.

Aliases are delimited by explicit ESC characters. The option `InputAutoReplacements` allows you to specify that certain kinds of input sequences should be immediately replaced even when they have no explicit delimiters. By default, for example, `->` is immediately replaced by →. You can give a rule of the form `"seq"->"rhs"` to specify that whenever *seq* appears as a token in your input, it should immediately be replaced by *rhs*.

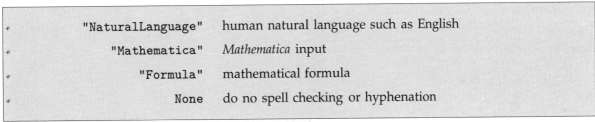

⊹	`"NaturalLanguage"`	human natural language such as English
⊹	`"Mathematica"`	*Mathematica* input
⊹	`"Formula"`	mathematical formula
⊹	`None`	do no spell checking or hyphenation

Settings for `LanguageCategory` to control spell checking and hyphenation.

The option `LanguageCategory` allows you to tell *Mathematica* what type of contents it should assume cells have. This determines how spelling and structure should be checked, and how hyphenation should be done.

option	typical default value	
`StructuredSelection`	False	whether to allow only complete subexpressions to be selected
`DragAndDrop`	False	whether to allow drag-and-drop editing

Options associated with interactive manipulation of expressions.

Mathematica normally allows you to select any part of an expression that you see on the screen. Occasionally, however, you may find it useful to get *Mathematica* to allow only selections which correspond to complete subexpressions. You can do this by setting the option StructuredSelection->True.

Here is an expression with a piece selected.

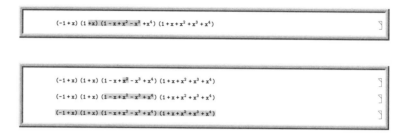

With StructuredSelection->True only complete subexpressions can ever be selected.

GridBox[*data*, *opts*]	give options that apply to a particular grid box
StyleBox[*boxes*, *opts*]	give options that apply to all boxes in *boxes*
Cell[*contents*, *opts*]	give options that apply to all boxes in *contents*
Cell[*contents*, GridBoxOptions->*opts*]	
	give default options settings for all GridBox objects in *contents*

Examples of specifying options for the display of expressions.

As discussed in Section 2.8 of the complete *Mathematica Book*, *Mathematica* provides many options for specifying how expressions should be displayed. By using StyleBox[*boxes*, *opts*] you can apply such options to collections of boxes. But *Mathematica* is set up so that any option that you can give to a StyleBox can also be given to a complete Cell object, or even a complete Notebook. Thus, for example, options like Background and LineIndent can be given to complete cells as well as to individual StyleBox objects.

There are some options that apply only to a particular type of box, such as GridBox. Usually these options are best given separately in each GridBox where they are needed. But sometimes you may want to specify default settings to be inherited by all GridBox objects that appear in a particular cell. You can do this by giving these default settings as the value of the option GridBoxOptions for the whole cell.

For each box type named *XXX*Box, *Mathematica* provides a cell option *XXX*BoxOptions that allows you to specify the default options settings for that type of box.

2.11 Files and Streams

■ 2.11.7 Importing and Exporting Files

Import["*file*", "List"]	import a one-dimensional list of data from a file
Export["*file*", *list*, "List"]	export *list* to a file as a one-dimensional list of data
Import["*file*", "Table"]	import a two-dimensional table of data from a file
Export["*file*", *list*, "Table"]	export *list* to a file as a two-dimensional table of data

Importing and exporting lists and tables of data.

This exports a list of data to the file out1.dat.

```
In[1]:= Export["out1.dat", {6.7, 8.2, -5.3}, "List"]
Out[1]= out1.dat
```

Here are the contents of the file.

```
In[2]:= !!out1.dat
6.7
8.2
-5.3
```

This imports the contents back into *Mathematica*.

```
In[3]:= Import["out1.dat", "List"]
Out[3]= {6.7, 8.2, -5.3}
```

If you want to use data purely within *Mathematica*, then the best way to keep it in a file is usually as a complete *Mathematica* expression, with all its structure preserved, as discussed on page 598 of the complete *Mathematica Book*. But if you want to exchange data with other programs, it is often more convenient to have the data in a simple list or table format.

This exports a two-dimensional array of data.

```
In[4]:= Export["out2.dat", {{5.6 10^12, 7.2 10^12}, {3, 5}}, "Table"]
Out[4]= out2.dat
```

When necessary, numbers are written in C or Fortran-like "E" notation.

```
In[5]:= !!out2.dat
5.6e12    7.2e12
3         5
```

This imports the array back into *Mathematica*.

```
In[6]:= Import["out2.dat", "Table"]
```
$$Out[6]= \{\{5.6 \times 10^{12}, 7.2 \times 10^{12}\}, \{3, 5\}\}$$

If you have a file in which each line consists of a single number, then you can use Import["*file*", "List"] to import the contents of the file as a list of numbers. If each line consists of a sequence of numbers separated by tabs or spaces, then Import["*file*", "Table"] will yield a list of lists of numbers. If the file contains items that are not numbers, then these are returned as *Mathematica* strings.

This exports a mixture of textual and numerical data.

```
In[7]:= Export["out3.dat", {{"first", 3.4}, {"second", 7.8}}]
Out[7]= out3.dat
```

Here is the exported data.

```
In[8]:= !!out3.dat
first    3.4
second   7.8
```

This imports the data back into *Mathematica*.

```
In[9]:= Import["out3.dat", "Table"]
Out[9]= {{first, 3.4}, {second, 7.8}}
```

With InputForm, you can explicitly see the strings.

```
In[10]:= InputForm[%]
Out[10]//InputForm= {{"first", 3.4}, {"second", 7.8}}
```

✦	Import["*file*", "List"]	treat each line as a separate numerical or other data item
✦	Import["*file*", "Table"]	treat each element on each line as a separate numerical or other data item
✦	Import["*file*", "Text"]	treat the whole file as a single string of text
✦	Import["*file*", "Lines"]	treat each line as a string of text
✦	Import["*file*", "Words"]	treat each separated word as a string of text

Importing files in different formats.

This creates a file with two lines of text.

```
In[11]:= Export["out4.dat",
            {"The first line.", "The second line."}, "Lines"]
Out[11]= out4.dat
```

Here are the contents of the file.

```
In[12]:= !!out4.dat
The first line.
The second line.
```

This imports the whole file as a single string.

```
In[13]:= Import["out4.dat", "Text"]//InputForm
Out[13]//InputForm= "The first line.\nThe second line.\n"
```

This imports the file as a list of lines of text.

```
In[14]:= Import["out4.dat", "Lines"]//InputForm
Out[14]//InputForm= {"The first line.", "The second line."}
```

This imports the file as a list of words separated by white space.

```
In[15]:= Import["out4.dat", "Words"]//InputForm
Out[15]//InputForm=
        {"The", "first", "line.", "The", "second", "line."}
```

3.2 Mathematical Functions

■ 3.2.4 Integer and Number-Theoretical Functions

Mod[k, n]	k modulo n (remainder from dividing k by n)
Quotient[m, n]	the quotient of m and n (integer part of m/n)
GCD[n_1, n_2, ...]	the greatest common divisor of n_1, n_2, ...
LCM[n_1, n_2, ...]	the least common multiple of n_1, n_2, ...
KroneckerDelta[n_1, n_2, ...]	the Kronecker delta $\delta_{n_1 n_2 \dots}$ equal to 1 if all the n_i are equal, and 0 otherwise
IntegerDigits[n, b]	the digits of n in base b
IntegerExponent[n, b]	the highest power of b that divides n

Some integer functions.

The remainder on dividing 17 by 3.	*In[1]:=* **Mod[17, 3]**
	Out[1]= 2
The integer part of 17/3.	*In[2]:=* **Quotient[17, 3]**
	Out[2]= 5
Mod also works with real numbers.	*In[3]:=* **Mod[5.6, 1.2]**
	Out[3]= 0.8
The result from Mod always has the same sign as the second argument.	*In[4]:=* **Mod[-5.6, 1.2]**
	Out[4]= 0.4

For any integers a and b, it is always true that $b*$Quotient[a, b] + Mod[a, b] is equal to a.

Mod[k, n]	result in the range 0 to $n - 1$
Mod[k, n, 1]	result in the range 1 to n
Mod[k, n, $-n/2$]	result in the range $\lceil -n/2 \rceil$ to $\lfloor +n/2 \rfloor$
Mod[k, n, d]	result in the range d to $d + n - 1$

Integer remainders with offsets.

Particularly when you are using Mod to get indices for parts of objects, you will often find it convenient to specify an offset.

This effectively extracts the 18th part of the list, with the list treated cyclically.	*In[5]:=* **Part[{a, b, c}, Mod[18, 3, 1]]**
	Out[5]= c

•

•

•

The **Kronecker delta** or **Kronecker symbol** KroneckerDelta[n_1, n_2, ...] is equal to 1 if all the n_i are equal, and is 0 otherwise. $\delta_{n_1 n_2...}$ can be thought of as a totally symmetric tensor.

This gives a totally symmetric tensor of rank 3.	*In[6]:=* **Array[KroneckerDelta, {3, 3, 3}]**
	Out[6]= {{{1, 0, 0}, {0, 0, 0}, {0, 0, 0}},
	{{0, 0, 0}, {0, 1, 0}, {0, 0, 0}},
	{{0, 0, 0}, {0, 0, 0}, {0, 0, 1}}}

•

•

•

PowerMod[a, b, n]	the power a^b modulo n
EulerPhi[n]	the Euler totient function $\phi(n)$
MoebiusMu[n]	the Möbius function $\mu(n)$
DivisorSigma[k, n]	the divisor function $\sigma_k(n)$
JacobiSymbol[n, m]	the Jacobi symbol $\left(\frac{n}{m}\right)$
ExtendedGCD[m, n]	the extended gcd of m and n
⚡ MultiplicativeOrder[k, n]	the multiplicative order of k modulo n
⚡ MultiplicativeOrder[k, n, {r_1, r_2, ... }]	
	the generalized multiplicative order with residues r_i
⚡ CarmichaelLambda[n]	the Carmichael function $\lambda(n)$
LatticeReduce[{v_1, v_2, ... }]	the reduced lattice basis for the set of integer vectors v_i

Some functions from number theory.

•
•
•

The **multiplicative order function** `MultiplicativeOrder[k, n]` gives the smallest integer m such that $k^m \equiv 1 \bmod n$. The function is sometimes known as the **index** or **discrete log** of k. The notation $\mathrm{ord}_n(k)$ is occasionally used.

The **generalized multiplicative order function** `MultiplicativeOrder[k, n, {r_1, r_2, ... }]` gives the smallest integer m such that $k^m \equiv r_i \bmod n$ for any i. `MultiplicativeOrder[k, n, {-1, 1}]` is sometimes known as the **suborder function** of k modulo n, denoted $\mathrm{sord}_n(k)$.

The **Carmichael function** or **least universal exponent** $\lambda(n)$ gives the smallest integer m such that $k^m \equiv 1 \bmod n$ for all integers k relatively prime to n.

•
•
•

⟋	`ContinuedFraction[x, n]`	generate the first n terms in the continued fraction representation of x
⟋	`FromContinuedFraction[list]`	reconstruct a number from its continued fraction representation

Continued fractions.

This generates the first 10 terms in the continued fraction representation for π.

```
In[7]:= ContinuedFraction[Pi, 10]
Out[7]= {3, 7, 15, 1, 292, 1, 1, 1, 2, 1}
```

This reconstructs the number represented by the list of continued fraction terms.

```
In[8]:= FromContinuedFraction[%]
```
$$Out[8]= \frac{1146408}{364913}$$

The result is close to π.

```
In[9]:= N[%]
Out[9]= 3.14159
```

Continued fractions appear in many number-theoretical settings. Rational numbers have terminating continued fraction representations. Quadratic irrational numbers have continued fraction representations that become repetitive.

✦	`ContinuedFraction[x]`	the complete continued fraction representation for a rational or quadratic irrational number
✦	`RealDigits[x]`	the complete digit sequence for a rational number
✦	`RealDigits[x, b]`	the complete digit sequence in base *b*

Complete representations for numbers.

The continued fraction representation of $\sqrt{79}$ starts with the term 8, then involves a sequence of terms that repeat forever.

```
In[10]:= ContinuedFraction[Sqrt[79]]
Out[10]= {8, {1, 7, 1, 16}}
```

This reconstructs $\sqrt{79}$ from its continued fraction representation.

```
In[11]:= FromContinuedFraction[%]
Out[11]= √79
```

This shows the recurring sequence of decimal digits in 3/7.

```
In[12]:= RealDigits[3/7]
Out[12]= {{{4, 2, 8, 5, 7, 1}}, 0}
```

FromDigits reconstructs the original number.

```
In[13]:= FromDigits[%]
Out[13]= 3/7
```

✦	`DigitCount[n, b, d]`	the number of *d* digits in the base *b* representation of *n*

Digit count function.

Here are the digits in the base 2 representation of the number 77.

```
In[14]:= IntegerDigits[77, 2]
Out[14]= {1, 0, 0, 1, 1, 0, 1}
```

This directly computes the number of ones in the base 2 representation.

```
In[15]:= DigitCount[77, 2, 1]
Out[15]= 4
```

The plot of the digit count function is self-similar.

$In[16]:=$ `ListPlot[Table[DigitCount[n, 2, 1], {n, 128}],`
              `PlotJoined->True]`

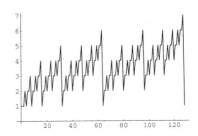

$\stackrel{\centerdot}{}$	`BitAnd[`n_1`,` n_2`, ... ]`	bitwise AND of the integers n_i
$\stackrel{\centerdot}{}$	`BitOr[`n_1`,` n_2`, ... ]`	bitwise OR of the integers n_i
$\stackrel{\centerdot}{}$	`BitXor[`n_1`,` n_2`, ... ]`	bitwise XOR of the integers n_i
$\stackrel{\centerdot}{}$	`BitNot[`n`]`	bitwise NOT of the integer n

Bitwise operations.

Bitwise operations act on integers represented as binary bits. `BitAnd[`n_1`,` n_2`, ... ]` yields the integer whose binary bit representation has ones at positions where the binary bit representations of all of the n_i have ones. `BitOr[`n_1`,` n_2`, ... ]` yields the integer with ones at positions where any of the n_i have ones. `BitXor[`n_1`,` n_2`]` yields the integer with ones at positions where n_1 or n_2 but not both have ones. `BitXor[`n_1`,` n_2`, ... ]` has ones where an odd number of the n_i have ones.

This finds the bitwise AND of the numbers 23 and 29 entered in base 2.

$In[17]:=$ `BaseForm[BitAnd[2^^10111, 2^^11101], 2]`

$Out[17]//BaseForm=$ 10101_2

Bitwise operations are used in various combinatorial algorithms. They are also commonly used in manipulating bitfields in low-level computer languages. In such languages, however, integers normally have a limited number of digits, typically a multiple of 8. Bitwise operations in *Mathematica* in effect allow integers to have an unlimited number of digits. When an integer is negative, it is taken to be represented in two's complement form, with an infinite sequence of ones on the left. This allows `BitNot[`n`]` to be equivalent simply to $-1 - n$.

■ 3.2.5 Combinatorial Functions

$n!$	factorial $n(n-1)(n-2) \times \ldots \times 1$
$n!!$	double factorial $n(n-2)(n-4) \times \ldots$
Binomial[n, m]	binomial coefficient $\binom{n}{m} = n!/[m!(n-m)!]$
Multinomial[n_1, n_2, ...]	multinomial coefficient $(n_1 + n_2 + \ldots)!/(n_1! n_2! \ldots)$
Fibonacci[n]	Fibonacci number F_n
Fibonacci[n, x]	Fibonacci polynomial $F_n(x)$
HarmonicNumber[n]	harmonic number H_n
HarmonicNumber[n, r]	harmonic number $H_n^{(r)}$ of order r
BernoulliB[n]	Bernoulli number B_n
BernoulliB[n, x]	Bernoulli polynomial $B_n(x)$
EulerE[n]	Euler number E_n
EulerE[n, x]	Euler polynomial $E_n(x)$
StirlingS1[n, m]	Stirling number of the first kind $S_n^{(m)}$
StirlingS2[n, m]	Stirling number of the second kind $\mathcal{S}_n^{(m)}$
PartitionsP[n]	the number $p(n)$ of unrestricted partitions of the integer n
PartitionsQ[n]	the number $q(n)$ of partitions of n into distinct parts
Signature[{i_1, i_2, ... }]	the signature of a permutation

Combinatorial functions.

•
•
•

The **harmonic numbers** HarmonicNumber[n] are given by $H_n = \sum_{i=1}^{n} 1/i$; the harmonic numbers of order r HarmonicNumber[n, r] are given by $H_n^{(r)} = \sum_{i=1}^{n} 1/i^r$. Harmonic numbers appear in many combinatorial estimation problems, often playing the role of discrete analogs of logarithms.

$\bullet$

$\bullet$

$\bullet$

■ 3.2.8 Mathematical Constants

I	$i = \sqrt{-1}$
Infinity	∞
Pi	$\pi \simeq 3.14159$
Degree	$\pi/180$: degrees to radians conversion factor
GoldenRatio	$\phi = (1 + \sqrt{5})/2 \simeq 1.61803$
E	$e \simeq 2.71828$
EulerGamma	Euler's constant $\gamma \simeq 0.577216$
Catalan	Catalan's constant $\simeq 0.915966$
Khinchin	Khinchin's constant $\simeq 2.68545$
Glaisher	Glaisher's constant $\simeq 1.28243$

Mathematical constants.

Euler's constant EulerGamma is given by the limit $\gamma = \lim_{m \to \infty} \left(\sum_{k=1}^{m} \frac{1}{k} - \log m \right)$. It appears in many integrals, and asymptotic formulas. It is sometimes known as the **Euler-Mascheroni constant**, and denoted C.

Catalan's constant Catalan is given by the sum $\sum_{k=0}^{\infty} (-1)^k (2k+1)^{-2}$. It often appears in asymptotic estimates of combinatorial functions.

Khinchin's constant (sometimes called Khintchine's constant) is given by $\prod_{s=1}^{\infty} (1 + \frac{1}{s(s+2)})^{\log_2 s}$. It gives the geometric mean of the terms in the continued fraction representation for a typical real number.

Glaisher's constant Glaisher A (sometimes called the Glaisher-Kinkelin constant) satisfies $\log(A) = \frac{1}{12} - \zeta'(-1)$, where ζ is the Riemann zeta function. It appears in various sums and integrals, particularly those involving gamma and zeta functions.

Mathematical constants can be evaluated to arbitrary precision.	`In[1]:= N[EulerGamma, 40]`
	`Out[1]= 0.5772156649015328606065120900824024310421`

Exact computations can also be done with them.	`In[2]:= IntegerPart[GoldenRatio^100]`
	`Out[2]= 792070839848372253126`

■ 3.2.10 Special Functions

Zeta and Related Functions

`LerchPhi[z, s, a]`	Lerch's transcendent $\Phi(z, s, a)$
`PolyLog[n, z]`	polylogarithm function $\mathrm{Li}_n(z)$
`PolyLog[n, p, z]`	Nielsen generalized polylogarithm function $S_{n,p}(z)$
`RiemannSiegelTheta[t]`	Riemann-Siegel function $\vartheta(t)$
`RiemannSiegelZ[t]`	Riemann-Siegel function $Z(t)$
`StieltjesGamma[n]`	Stieltjes constants γ_n
`Zeta[s]`	Riemann zeta function $\zeta(s)$
`Zeta[s, a]`	generalized Riemann zeta function $\zeta(s, a)$

Zeta and related functions.

•
•
•

The **polylogarithm functions** `PolyLog[n, z]` are given by $\mathrm{Li}_n(z) = \sum_{k=1}^{\infty} z^k/k^n$. The polylogarithm function is sometimes known as **Jonquière's function**. The **dilogarithm** `PolyLog[2, z]` satisfies $\mathrm{Li}_2(z) = \int_z^0 \log(1 - t)/t\, dt$. Sometimes $\mathrm{Li}_2(1 - z)$ is known as **Spence's integral**. The **Nielsen generalized polylogarithm functions** or **hyperlogarithms** `PolyLog[n, p, z]` are given by $S_{n,p}(z) = (-1)^{n+p-1}/((n-1)!\,p!) \int_0^1 \log^{n-1}(t) \log^p(1 - zt)/t\, dt$. Polylogarithm functions appear in Feynman diagram integrals in elementary particle physics, as well as in algebraic K-theory.

- •
- •
- •

Bessel and Related Functions

AiryAi[z] and AiryBi[z]	Airy functions Ai(z) and Bi(z)
AiryAiPrime[z] and AiryBiPrime[z]	derivatives of Airy functions Ai′(z) and Bi′(z)
BesselJ[n, z] and BesselY[n, z]	Bessel functions $J_n(z)$ and $Y_n(z)$
BesselI[n, z] and BesselK[n, z]	modified Bessel functions $I_n(z)$ and $K_n(z)$
StruveH[n, z] and StruveL[n, z]	Struve function $\mathbf{H}_n(z)$ and modified Struve function $\mathbf{L}_n(z)$

Bessel and related functions.

- •
- •
- •

The **Struve function** StruveH[n, z] appears in the solution of the inhomogeneous Bessel equation which for integer n has the form $z^2 y'' + zy' + (z^2 - n^2)y = \frac{2}{\pi}\frac{z^{n+1}}{(2n-1)!!}$; the general solution to this equation consists of a linear combination of Bessel functions with the Struve function $\mathbf{H}_n(z)$ added. The **modified Struve function** StruveL[n, z] is given in terms of the ordinary Struve function by $\mathbf{L}_n(z) = -ie^{-in\pi/2}\mathbf{H}_n(z)$. Struve functions appear particularly in electromagnetic theory.

- •
- •
- •

⁓ Hypergeometric Functions and Generalizations

Hypergeometric2F1[a, b, c, z]   hypergeometric function $_2F_1(a, b; c; z)$

+ Hypergeometric2F1Regularized[a, b, c, z]

        regularized hypergeometric function $_2F_1(a, b; c; z)/\Gamma(c)$

+ HypergeometricPFQ[$\{a_1, \ldots, a_p\}$, $\{b_1, \ldots, b_q\}$, z]

        generalized hypergeometric function $_pF_q(\mathbf{a}; \mathbf{b}; z)$

+ HypergeometricPFQRegularized[$\{a_1, \ldots, a_p\}$, $\{b_1, \ldots, b_q\}$, z]

        regularized generalized hypergeometric function

+ MeijerG[$\{\{a_1, \ldots, a_n\}, \{a_{n+1}, \ldots, a_p\}\}$, $\{\{b_1, \ldots, b_m\}, \{b_{m+1}, \ldots, b_q\}\}$, z]

        Meijer G function

+ AppellF1[a, b_1, b_2, c, x, y]   Appell hypergeometric function of two variables

        $F_1(a; b_1, b_2; c; x, y)$

Hypergeometric functions and generalizations.

•

•

•

The **generalized hypergeometric function** or **Barnes extended hypergeometric function** HypergeometricPFQ[$\{a_1, \ldots, a_p\}$, $\{b_1, \ldots, b_q\}$, z] has series expansion $_pF_q(\mathbf{a}; \mathbf{b}; z) = \sum_{k=0}^{\infty} (a_1)_k \ldots (a_p)_k / [(b_1)_k \ldots (b_q)_k] z^k / k!$.

The **Meijer G function** MeijerG[$\{\{a_1, \ldots, a_n\}, \{a_{n+1}, \ldots, a_p\}\}$, $\{\{b_1, \ldots, b_m\}, \{b_{m+1}, \ldots, b_q\}\}$, z] is defined by the contour integral representation $G_{pq}^{mn}\left(z \left|\begin{smallmatrix} a_1, \ldots, a_p \\ b_1, \ldots, b_q \end{smallmatrix}\right.\right) = \frac{1}{2\pi i} \int \Gamma(1 - a_1 - s) \ldots \Gamma(1 - a_n - s) \times \Gamma(b_1 + s) \ldots \Gamma(b_m + s) / (\Gamma(a_{n+1} + s) \ldots \Gamma(a_p + s) \Gamma(1 - b_{m+1} - s) \ldots \Gamma(1 - b_q - s)) z^{-s} ds$, where the contour of integration is set up to lie between the poles of $\Gamma(1 - a_i - s)$ and the poles of $\Gamma(b_i + s)$. MeijerG is a very general function whose special cases cover most of the functions discussed in the past few sections.

The **Appell hypergeometric function of two variables** AppellF1[a, b_1, b_2, c, x, y] has series expansion $F_1(a; b_1, b_2; c; x, y) = \sum_{m=0}^{\infty} \sum_{n=0}^{\infty} (a)_{m+n} (b_1)_m (b_2)_n / (m! n! (c)_{m+n}) x^m y^n$. This function appears for example in integrating cubic polynomials to arbitrary powers.

3.3 Algebraic Manipulation

■ 3.3.4 Algebraic Operations on Polynomials

For many kinds of practical calculations, the only operations you will need to perform on polynomials are essentially the structural ones discussed in the preceding sections.

If you do more advanced algebra with polynomials, however, you will have to use the algebraic operations discussed in this section.

You should realize that most of the operations discussed in this section work only on ordinary polynomials, with integer exponents and rational-number coefficients for each term.

PolynomialQuotient[$poly_1$, $poly_2$, x]	find the result of dividing the polynomial $poly_1$ in x by $poly_2$, dropping any remainder term
PolynomialRemainder[$poly_1$, $poly_2$, x]	find the remainder from dividing the polynomial $poly_1$ in x by $poly_2$
PolynomialGCD[$poly_1$, $poly_2$]	find the greatest common divisor of two polynomials
PolynomialLCM[$poly_1$, $poly_2$]	find the least common multiple of two polynomials
PolynomialMod[$poly$, m]	reduce the polynomial $poly$ modulo m
Resultant[$poly_1$, $poly_2$, x]	find the resultant of two polynomials
*Subresultants[$poly_1$, $poly_2$, x]	find the principal subresultant coefficients of two polynomials
GroebnerBasis[{$poly_1$, $poly_2$, ... }, {x_1, x_2, ... }]	find the Gröbner basis for the polynomials $poly_i$
GroebnerBasis[{$poly_1$, $poly_2$, ... }, {x_1, x_2, ... }, {y_1, y_2, ... }]	find the Gröbner basis eliminating the y_i
+ PolynomialReduce[$poly$, {$poly_1$, $poly_2$, ... }, {x_1, x_2, ... }]	find a minimal representation of $poly$ in terms of the $poly_i$

Reduction of polynomials.

•
•
•

The function Resultant[$poly_1$, $poly_2$, x] is used in a number of classical algebraic algorithms. The resultant of two polynomials a and b, both with leading coefficient one, is given by the product of all the differences $a_i - b_j$ between the roots of the polynomials. It turns out that for any pair of polynomials, the resultant is always a polynomial in their coefficients. By looking at when the resultant is zero, one can tell for what values of their parameters two polynomials have a common root. Two polynomials with leading coefficient one have k common roots if exactly the first k elements in the list Subresultants[$poly_1$, $poly_2$, x] are zero.

•
•
•

■ 3.3.9 Simplification

Simplify[$expr$]	try various algebraic and trigonometric transformations to simplify an expression
FullSimplify[$expr$]	try a much wider range of transformations

Simplifying expressions.

Mathematica does not automatically simplify an algebraic expression like this.

In[1]:= (1 - x)/(1 - x^2)

Out[1]= $\dfrac{1-x}{1-x^2}$

Simplify performs the simplification.

In[2]:= Simplify[%]

Out[2]= $\dfrac{1}{1+x}$

Simplify performs standard algebraic and trigonometric simplifications.

In[3]:= Simplify[Sin[x]^2 + Cos[x]^2]

Out[3]= 1

It does not, however, do more sophisticated transformations that involve, for example, special functions.

In[4]:= Simplify[Gamma[1+n]/n]

Out[4]= $\dfrac{\text{Gamma}[1+n]}{n}$

FullSimplify does perform such transformations.

```
In[5]:= FullSimplify[%]
Out[5]= Gamma[n]
```

+ `FullSimplify[`*expr*`, ExcludedForms -> `*pattern*`]`
   try to simplify *expr*, without touching subexpressions that match *pattern*

Controlling simplification.

Here is an expression involving trigonometric functions and square roots.

```
In[6]:= t = (1 - Sin[x]^2) Sqrt[Expand[(1 + Sqrt[2])^20]]
```
$$Out[6]= \sqrt{22619537 + 15994428\sqrt{2}}\ \left(1 - \text{Sin}[x]^2\right)$$

By default, FullSimplify will try to simplify everything.

```
In[7]:= FullSimplify[t]
```
$$Out[7]= \left(3363 + 2378\sqrt{2}\right)\text{Cos}[x]^2$$

This makes FullSimplify avoid simplifying the square roots.

```
In[8]:= FullSimplify[t, ExcludedForms->Sqrt[_]]
```
$$Out[8]= \sqrt{22619537 + 15994428\sqrt{2}}\ \text{Cos}[x]^2$$

+ `FullSimplify[`*expr*`, TimeConstraint->`*t*`]`
   try to simplify *expr*, working for at most *t* seconds on each transformation

+ `FullSimplify[`*expr*`, TransformationFunctions -> {`f_1, f_2, ... `}]`
   use only the functions f_i in trying to transform parts of *expr*

+ `FullSimplify[`*expr*`, TransformationFunctions -> {Automatic, `f_1, f_2, ... `}]`
   use built-in transformations as well as the f_i

+ `Simplify[`*expr*`, ComplexityFunction->`*c*`]` and `FullSimplify[`*expr*`, ComplexityFunction->`*c*`]`
   simplify using *c* to determine what form is considered simplest

Further control of simplification.

In both `Simplify` and `FullSimplify` there is always an issue of what counts as the "simplest" form of an expression. You can use the option `ComplexityFunction -> `*c* to provide a function to determine this. The function will be applied to each candidate form of the expression, and the one that gives the smallest numerical value will be considered simplest.

With its default definition of simplicity, Simplify leaves this unchanged.

```
In[9]:= Simplify[4 Log[10]]
Out[9]= 4 Log[10]
```

This now tries to minimize the number of elements in the expression.

```
In[10]:= Simplify[4 Log[10], ComplexityFunction -> LeafCount]
Out[10]= Log[10000]
```

■ 3.3.10 Using Assumptions

Mathematica normally makes as few assumptions as possible about the objects you ask it to manipulate. This means that the results it gives are as general as possible. But sometimes these results are considerably more complicated than they would be if more assumptions were made.

Simplify[*expr*, *assum*]	simplify with assumptions
FullSimplify[*expr*, *assum*]	full simplify with assumptions
FunctionExpand[*expr*, *assum*]	function expand with assumptions

Doing operations with assumptions.

Simplify by default does essentially nothing with this expression.

```
In[1]:= Simplify[1/Sqrt[x] - Sqrt[1/x]]
```
$$Out[1]= -\sqrt{\frac{1}{x}} + \frac{1}{\sqrt{x}}$$

The reason is that its value is quite different for different choices of x.

```
In[2]:= % /. x -> {-3, -2, -1, 1, 2, 3}
```
$$Out[2]= \left\{ -\frac{2\,i}{\sqrt{3}}, -i\sqrt{2}, -2\,i, 0, 0, 0 \right\}$$

With the assumption $x > 0$, Simplify can immediately reduce the expression to 0.

```
In[3]:= Simplify[1/Sqrt[x] - Sqrt[1/x], x > 0]
Out[3]= 0
```

Without making assumptions about x and y, nothing can be done.

```
In[4]:= FunctionExpand[Log[x y]]
Out[4]= Log[x y]
```

If x and y are both assumed positive, the log can be expanded.

```
In[5]:= FunctionExpand[Log[x y], x > 0 && y > 0]
Out[5]= Log[x] + Log[y]
```

By applying Simplify and FullSimplify with appropriate assumptions to equations and inequalities you can in effect establish a vast range of theorems.

Without making assumptions about x the truth or falsity of this equation cannot be determined.

```
In[6]:= Simplify[Abs[x] == x]
Out[6]= Abs[x] == x
```

Now Simplify can prove that the equation is true.

```
In[7]:= Simplify[Abs[x] == x, x > 0]
Out[7]= True
```

This establishes the standard result that the arithmetic mean is larger than the geometric one.

```
In[8]:= Simplify[(x + y)/2 >= Sqrt[x y], x >= 0 && y >= 0]
Out[8]= True
```

This proves that erf(x) lies in the range $(0, 1)$ for all positive arguments.

```
In[9]:= FullSimplify[0 < Erf[x] < 1, x > 0]
Out[9]= True
```

An important class of assumptions are those which assert that some object is an element of a particular domain. You can set up such assumptions using $x \in dom$, where the $\in$ character can be entered as :elem: or \[Element].

⊬	$x \in dom$ or Element[x, dom]	assert that x is an element of the domain *dom*
⊬	$\{x_1, x_2, \dots\} \in dom$	assert that all the x_i are elements of the domain *dom*
⊬	$patt \in dom$	assert that any expression which matches *patt* is an element of the domain *dom*

Asserting that objects are elements of domains.

This confirms that π is an element of the domain of real numbers.

```
In[10]:= Pi ∈ Reals
Out[10]= True
```

These numbers are all elements of the domain of algebraic numbers.

```
In[11]:= {1, Sqrt[2], 3 + Sqrt[5]} ∈ Algebraics
Out[11]= True
```

Mathematica knows that π is not an algebraic number.

```
In[12]:= Pi ∈ Algebraics
Out[12]= False
```

Current mathematics has not established whether $e + \pi$ is an algebraic number or not.

```
In[13]:= E + Pi ∈ Algebraics
Out[13]= e + π ∈ Algebraics
```

This represents the assertion that the symbol x is an element of the domain of real numbers.

```
In[14]:= x ∈ Reals
Out[14]= x ∈ Reals
```

✳	Complexes	the domain of complex numbers $\mathbb{C}$
✳	Reals	the domain of real numbers $\mathbb{R}$
✳	Algebraics	the domain of algebraic numbers $\mathbb{A}$
✳	Rationals	the domain of rational numbers $\mathbb{Q}$
✳	Integers	the domain of integers $\mathbb{Z}$
✳	Primes	the domain of primes $\mathbb{P}$
✳	Booleans	the domain of booleans (True and False) $\mathbb{B}$

Domains supported by *Mathematica*.

If n is assumed to be an integer, $\sin(n\pi)$ is zero.

```
In[15]:= Simplify[Sin[n Pi], n ∈ Integers]
Out[15]= 0
```

This establishes the theorem $\cosh(x) \geq 1$ if x is assumed to be a real number.

```
In[16]:= Simplify[Cosh[x] >= 1, x ∈ Reals]
Out[16]= True
```

If you say that a variable satisfies an inequality, *Mathematica* will automatically assume that it is real.

```
In[17]:= Simplify[x ∈ Reals, x > 0]
Out[17]= True
```

By using `Simplify`, `FullSimplify` and `FunctionExpand` with assumptions you can access many of *Mathematica*'s vast collection of mathematical facts.

This uses the periodicity of the tangent function.

```
In[18]:= Simplify[Tan[x + Pi k], k ∈ Integers]
Out[18]= Tan[x]
```

The assumption k/2 ∈ Integers implies that k must be even.

```
In[19]:= Simplify[Tan[x + Pi k/2], k/2 ∈ Integers]
Out[19]= Tan[x]
```

Mathematica knows that $\log(x) < \exp(x)$ for positive x.

```
In[20]:= Simplify[Log[x] < Exp[x], x > 0]
Out[20]= True
```

FullSimplify accesses knowledge about special functions.

```
In[21]:= FullSimplify[Im[BesselJ[0, x]], x ∈ Reals]
Out[21]= 0
```

Mathematica knows about discrete mathematics and number theory as well as continuous mathematics.

This uses Wilson's Theorem to simplify the result.

```
In[22]:= FunctionExpand[Mod[(p - 1)!, p], p ∈ Primes]
Out[22]= -1 + p
```

This uses the multiplicative property of the Euler phi function.

```
In[23]:= FunctionExpand[EulerPhi[m n], {m, n} ∈ Integers &&
                                              GCD[m, n] == 1]

Out[23]= EulerPhi[m] EulerPhi[n]
```

3.5 Calculus

▪ 3.5.11 Integral Transforms and Related Operations

Laplace Transforms

LaplaceTransform[*expr*, *t*, *s*]	the Laplace transform of *expr*
InverseLaplaceTransform[*expr*, *s*, *t*]	the inverse Laplace transform of *expr*

One-dimensional Laplace transforms.

The Laplace transform of a function $f(t)$ is given by $\int_0^\infty f(t)e^{-st}\,dt$. The inverse Laplace transform of $F(s)$ is given for suitable γ by $\frac{1}{2\pi i}\int_{\gamma-i\infty}^{\gamma+i\infty} F(s)e^{st}\,ds$.

Here is a simple Laplace transform.

In[1]:= **LaplaceTransform[t^4 Sin[t], t, s]**

$$Out[1]= \frac{24\,(1 - 10\,s^2 + 5\,s^4)}{(1 + s^2)^5}$$

Here is the inverse.

In[2]:= **InverseLaplaceTransform[%, s, t]**

Out[2]= t^4 Sin[t]

Even simple transforms often involve special functions.

In[3]:= **LaplaceTransform[1/(1 + t^2), t, s]**

$$Out[3]= \frac{1}{2}\,(2\,\text{CosIntegral}[s]\,\text{Sin}[s] + \text{Cos}[s]\,(\pi - 2\,\text{SinIntegral}[s]))$$

Here the result involves a Meijer G function.

In[4]:= **LaplaceTransform[1/(1 + t^3), t, s]**

$$Out[4]= \frac{\text{MeijerG}\!\left[\{\{\tfrac{2}{3}\},\,\{\}\},\,\{\{0,\,\tfrac{1}{3},\,\tfrac{2}{3},\,\tfrac{2}{3}\},\,\{\}\},\,\tfrac{s^3}{27}\right]}{2\sqrt{3}\,\pi}$$

The Laplace transform of a Bessel function involves a hypergeometric function.

In[5]:= **LaplaceTransform[BesselJ[n, t], t, s]**

$$Out[5]= \frac{2^{-n}\left(\tfrac{1}{s^2}\right)^{n/2}\text{Hypergeometric2F1}\!\left[\tfrac{1+n}{2},\,\tfrac{2+n}{2},\,1+n,\,-\tfrac{1}{s^2}\right]}{s}$$

Laplace transforms have the property that they turn integration and differentiation into essentially algebraic operations. They are therefore commonly used in studying systems governed by differential equations.

Integration becomes multiplication by $1/s$ when one does a Laplace transform.

```
In[6]:= LaplaceTransform[Integrate[f[u], {u, 0, t}], t, s]
```

$$Out[6]= \frac{\text{LaplaceTransform}[f[t], t, s]}{s}$$

+ LaplaceTransform[*expr*, {t_1, t_2, ... }, {s_1, s_2, ... }]
  the multidimensional Laplace transform of *expr*

+ InverseLaplaceTransform[*expr*, {s_1, s_2, ... }, {t_1, t_2, ... }]
  the multidimensional inverse Laplace transform of *expr*

Multidimensional Laplace transforms.

+ Fourier Transforms

+ FourierTransform[*expr*, *t*, ω]   the Fourier transform of *expr*

+ InverseFourierTransform[*expr*, ω, *t*]
  the inverse Fourier transform of *expr*

One-dimensional Fourier transforms.

Here is a Fourier transform.

```
In[7]:= FourierTransform[1/(1 + t^4), t, ω]
```

$$Out[7]= \left(\frac{1}{4} + \frac{\text{i}}{4}\right) \text{e}^{-\frac{(1+\text{i})\, \text{Abs}[\omega]}{\sqrt{2}}} \left(1 - \text{i}\, \text{e}^{\text{i}\, \sqrt{2}\, \text{Abs}[\omega]}\right) \sqrt{\pi}$$

This finds the inverse.

```
In[8]:= InverseFourierTransform[%, ω, t]
```

$$Out[8]= \frac{1}{1 + t^4}$$

In *Mathematica* the Fourier transform of a function $f(t)$ is by default defined to be $\frac{1}{\sqrt{2\pi}} \int_{-\infty}^{\infty} f(t)\, e^{i\omega t}\, dt$.
The inverse Fourier transform of $F(\omega)$ is similarly defined as $\frac{1}{\sqrt{2\pi}} \int_{-\infty}^{\infty} F(\omega)\, e^{-i\omega t}\, d\omega$.

In different scientific and technical fields different conventions are often used for defining Fourier transforms. The option FourierParameters in *Mathematica* allows you to choose any of these conventions you want.

common convention	setting	Fourier transform	inverse Fourier transform
Mathematica default	{0, 1}	$\frac{1}{\sqrt{2\pi}} \int_{-\infty}^{\infty} f(t)\, e^{i\omega t}\, dt$	$\frac{1}{\sqrt{2\pi}} \int_{-\infty}^{\infty} F(\omega)\, e^{-i\omega t}\, d\omega$
pure mathematics	{1, -1}	$\int_{-\infty}^{\infty} f(t)\, e^{-i\omega t}\, dt$	$\frac{1}{2\pi} \int_{-\infty}^{\infty} F(\omega)\, e^{i\omega t}\, d\omega$
classical physics	{-1, 1}	$\frac{1}{2\pi} \int_{-\infty}^{\infty} f(t)\, e^{i\omega t}\, dt$	$\int_{-\infty}^{\infty} F(\omega)\, e^{-i\omega t}\, d\omega$
modern physics	{0, 1}	$\frac{1}{\sqrt{2\pi}} \int_{-\infty}^{\infty} f(t)\, e^{i\omega t}\, dt$	$\frac{1}{\sqrt{2\pi}} \int_{-\infty}^{\infty} F(\omega)\, e^{-i\omega t}\, d\omega$
systems engineering	{1, -1}	$\int_{-\infty}^{\infty} f(t)\, e^{-i\omega t}\, dt$	$\frac{1}{2\pi} \int_{-\infty}^{\infty} F(\omega)\, e^{i\omega t}\, d\omega$
signal processing	{0, -2 Pi}	$\int_{-\infty}^{\infty} f(t)\, e^{-2\pi i\omega t}\, dt$	$\int_{-\infty}^{\infty} F(\omega)\, e^{2\pi i\omega t}\, d\omega$
general case	{a, b}	$\sqrt{\frac{|b|}{(2\pi)^{1-a}}} \int_{-\infty}^{\infty} f(t)\, e^{ib\omega t}\, dt$	$\sqrt{\frac{|b|}{(2\pi)^{1+a}}} \int_{-\infty}^{\infty} F(\omega)\, e^{-ib\omega t}\, d\omega$

Typical settings for `FourierParameters` with various conventions.

Here is a Fourier transform with the default choice of parameters.

```
In[9]:= FourierTransform[Exp[-t^2], t, ω]
```

$$Out[9]= \frac{e^{-\frac{\omega^2}{4}}}{\sqrt{2}}$$

Here is the same Fourier transform with the choice of parameters typically used in signal processing.

```
In[10]:= FourierTransform[Exp[-t^2], t, ω,
                FourierParameters->{0, -2 Pi}]
```

$$Out[10]= e^{-\pi^2 \omega^2} \sqrt{\pi}$$

⊹ `FourierSinTransform[`*expr, t, ω*`]`	
	Fourier sine transform
⊹ `FourierCosTransform[`*expr, t, ω*`]`	
	Fourier cosine transform
⊹ `InverseFourierSinTransform[`*expr, ω, t*`]`	
	inverse Fourier sine transform
⊹ `InverseFourierCosTransform[`*expr, ω, t*`]`	
	inverse Fourier cosine transform

Fourier sine and cosine transforms.

In some applications of Fourier transforms, it is convenient to avoid ever introducing complex exponentials. Fourier sine and cosine transforms correspond to integrating respectively with $\sin(\omega t)$ and $\cos(\omega t)$ instead of $\exp(i\omega t)$, and using limits 0 and ∞ rather than $-\infty$ and ∞.

Here are the Fourier sine and cosine transforms of e^{-t}.

```
In[11]:= {FourierSinTransform[Exp[-t], t, ω],
                FourierCosTransform[Exp[-t], t, ω]}
```

$$Out[11]= \left\{ \frac{\sqrt{\frac{2}{\pi}}\,\omega}{1+\omega^2}, \frac{\sqrt{\frac{2}{\pi}}}{1+\omega^2} \right\}$$

+ FourierTransform[*expr*, {t_1, t_2, ... }, {ω_1, ω_2, ... }]

   the multidimensional Fourier transform of *expr*

+ InverseFourierTransform[*expr*, {ω_1, ω_2, ... }, {t_1, t_2, ... }]

   the multidimensional inverse Fourier transform of *expr*

+ FourierSinTransform[*expr*, {t_1, t_2, ... }, {ω_1, ω_2, ... }],
FourierCosTransform[*expr*, {t_1, t_2, ... }, {ω_1, ω_2, ... }]

   the multidimensional sine and cosine Fourier transforms of *expr*

+ InverseFourierSinTransform[*expr*, {ω_1, ω_2, ... }, {t_1, t_2, ... }],
InverseFourierCosTransform[*expr*, {ω_1, ω_2, ... }, {t_1, t_2, ... }]

   the multidimensional inverse Fourier sine and cosine transforms of *expr*

Multidimensional Fourier transforms.

This evaluates a two-dimensional Fourier transform.

```
In[12]:= FourierTransform[(u v)^2 Exp[-u^2-v^2], {u, v}, {a, b}]
```

$$Out[12]= \frac{1}{32}\,(-2+a^2)\,(-2+b^2)\,e^{-\frac{a^2}{4}-\frac{b^2}{4}}$$

This inverts the transform.

```
In[13]:= InverseFourierTransform[%, {a, b}, {u, v}]
```

$$Out[13]= e^{-u^2-v^2}\,u^2\,v^2$$

+ Z Transforms

+   ZTransform[*expr*, n, z]   Z transform of *expr*

+InverseZTransform[*expr*, z, n]   inverse Z transform of *expr*

Z transforms.

The Z transform of a function $f(n)$ is given by $\sum_{n=0}^{\infty} f(n)z^{-n}$. The inverse Z transform of $F(z)$ is given by the contour integral $\frac{1}{2\pi i} \oint F(z)z^{n-1}\,dz$. Z transforms are effectively discrete analogs of Laplace transforms. They are widely used for solving difference equations, especially in digital signal processing and control theory. They can be thought of as producing generating functions, of the kind commonly used in combinatorics and number theory.

This computes the Z transform of 2^{-n}.

$In[14]:=$ **ZTransform[2^-n, n, z]**

$Out[14]=$ $\dfrac{2\,z}{-1+2\,z}$

Here is the inverse Z transform.

$In[15]:=$ **InverseZTransform[%, z, n]**

$Out[15]=$ $2^{-n}\,\text{UnitStep}[n]$

The generating function for $1/n!$ is an exponential function.

$In[16]:=$ **ZTransform[1/n!, n, z]**

$Out[16]=$ $\mathrm{e}^{\frac{1}{z}}$

■ 3.5.12 Generalized Functions and Related Objects

In many practical situations it is convenient to consider limits in which a fixed amount of something is concentrated into an infinitesimal region. Ordinary mathematical functions of the kind normally encountered in calculus cannot readily represent such limits. However, it is possible to introduce *generalized functions* or *distributions* which can represent these limits in integrals and other types of calculations.

DiracDelta[x]	Dirac delta function $\delta(x)$
UnitStep[x]	unit step function, equal to 0 for $x < 0$ and 1 for $x > 0$

Dirac delta and unit step functions.

Here is a function concentrated around $x = 0$.

$In[1]:=$ **Plot[Sqrt[50/Pi] Exp[-50 x^2], {x, -2, 2}, PlotRange->All]**

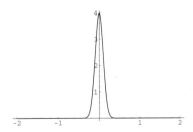

As n gets larger, the functions become progressively more concentrated.

```
In[2]:= Plot[Evaluate[Sqrt[n/Pi] Exp[-n x^2] /. n -> {1, 10, 100}],
            {x, -2, 2}, PlotRange->All];
```

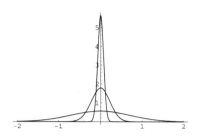

For any $n > 0$, their integrals are nevertheless always equal to 1.

```
In[3]:= Integrate[Sqrt[n/Pi] Exp[-n x^2], {x, -Infinity, Infinity},
            Assumptions -> n > 0]

Out[3]= 1
```

The limit of the functions for infinite n is effectively a Dirac delta function, whose integral is again 1.

```
In[4]:= Integrate[DiracDelta[x], {x, -Infinity, Infinity}]

Out[4]= 1
```

DiracDelta evaluates to 0 at all real points except $x = 0$.

```
In[5]:= Table[DiracDelta[x], {x, -3, 3}]

Out[5]= {0, 0, 0, DiracDelta[0], 0, 0, 0}
```

Inserting a delta function in an integral effectively causes the integrand to be sampled at discrete points where the argument of the delta function vanishes.

This samples the function f with argument 2.

```
In[6]:= Integrate[DiracDelta[x - 2] f[x], {x, -4, 4}]

Out[6]= f[2]
```

Here is a slightly more complicated example.

```
In[7]:= Integrate[DiracDelta[x^2 - x - 1], {x, 0, 2}]
```

$$Out[7]= \frac{1}{\sqrt{5}}$$

This effectively counts the number of zeros of $\cos(x)$ in the region of integration.

```
In[8]:= Integrate[DiracDelta[Cos[x]], {x, -30, 30}]

Out[8]= 20
```

The **unit step function** UnitStep[x] is effectively the indefinite integral of the delta function. It is sometimes known as the **Heaviside function**, and is variously denoted $H(x)$, $\theta(x)$, $\mu(x)$, and $U(x)$. It does not need to be considered as a generalized function, though it has a discontinuity at $x = 0$. The unit step function is often used in setting up piecewise continuous functions, and in representing signals and other quantities that become non-zero only beyond some point.

The indefinite integral of the delta function is the unit step function.

```
In[9]:= Integrate[DiracDelta[x], x]

Out[9]= UnitStep[x]
```

This generates a square wave.

In[10]:= **Plot[UnitStep[Sin[x]], {x, 0, 30}]**

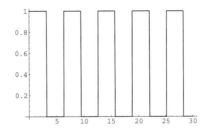

Here is the integral of the square wave.

In[11]:= **Integrate[UnitStep[Sin[x]], {x, 0, 30}]**

Out[11]= 5π

The value of the integral depends on whether a lies in the interval $(-2, 2)$.

In[12]:= **Integrate[f[x] DiracDelta[x - a], {x, -2, 2}]**

Out[12]= f[a] UnitStep[2 - a] UnitStep[2 + a]

DiracDelta and **UnitStep** often arise in doing integral transforms.

The Fourier transform of a constant function is a delta function.

In[13]:= **FourierTransform[1, t, ω]**

Out[13]= $\sqrt{2\pi}$ DiracDelta[ω]

The Fourier transform of cos(t) involves the sum of two delta functions.

In[14]:= **FourierTransform[Cos[t], t, ω]**

Out[14]= $\sqrt{\dfrac{\pi}{2}}$ (DiracDelta[-1 + ω] + DiracDelta[1 + ω])

Dirac delta functions can be used in **DSolve** to find the impulse response or Green's function of systems represented by linear and certain other differential equations.

This finds the behavior of a harmonic oscillator subjected to an impulse at $t = 0$.

In[15]:= **DSolve[{x''[t] + r x[t] == DiracDelta[t],**
              x[0]==0, x'[0]==1}, x[t], t]

Out[15]= $\left\{\left\{x[t] \rightarrow \dfrac{Sin\left[\sqrt{r}\ t\right] UnitStep[t]}{\sqrt{r}}\right\}\right\}$

+	DiracDelta[x_1, x_2, ...]	multidimensional Dirac delta function equal to 0 unless all the x_i are zero
+	UnitStep[x_1, x_2, ...]	multidimensional unit step function, equal to 0 unless all the x_i are positive

Multidimensional Dirac delta and unit step functions.

Related to the multidimensional Dirac delta function are two integer functions: discrete delta and Kronecker delta. Discrete delta $\delta(n_1, n_2, ...)$ is 1 if all the $n_i = 0$, and is zero otherwise. Kronecker delta $\delta_{n_1 n_2 ...}$ is 1 if all the n_i are equal, and is zero otherwise.

＋	`DiscreteDelta[`n_1`, `n_2`, ... ]`	discrete delta $\delta(n_1, n_2, ...)$
＋	`KroneckerDelta[`n_1`, `n_2`, ... ]`	Kronecker delta $\delta_{n_1 n_2 ...}$

Integer delta functions.

3.7 Linear Algebra

■ 3.7.2 Getting Pieces of Matrices

$m[[i, j]]$	the i,j^{th} entry
$m[[i]]$	the ith row
$m[[\text{All}, i]]$	the ith column
$\text{Take}[m, \{i_0, i_1\}, \{j_0, j_1\}]$	the submatrix with rows i_0 through i_1 and columns j_0 through j_1
$m[[\{i_1, \ldots, i_r\}, \{j_1, \ldots, j_s\}]]$	the $r \times s$ submatrix with elements having row indices i_k and column indices j_k
$\text{Tr}[m, \text{List}]$	elements on the diagonal

Ways to get pieces of matrices.

Matrices in *Mathematica* are represented as lists of lists. You can use all the standard *Mathematica* list-manipulation operations on matrices.

Here is a sample 3×3 matrix.

```
In[1]:= t = Array[a, {3, 3}]
Out[1]= {{a[1, 1], a[1, 2], a[1, 3]},
         {a[2, 1], a[2, 2], a[2, 3]},
         {a[3, 1], a[3, 2], a[3, 3]}}
```

This picks out the second row of the matrix.

```
In[2]:= t[[2]]
Out[2]= {a[2, 1], a[2, 2], a[2, 3]}
```

Here is the second column of the matrix.

```
In[3]:= t[[All, 2]]
Out[3]= {a[1, 2], a[2, 2], a[3, 2]}
```

This picks out a submatrix.

```
In[4]:= Take[t, {1, 2}, {2, 3}]
Out[4]= {{a[1, 2], a[1, 3]}, {a[2, 2], a[2, 3]}}
```

■ 3.7.7 Basic Matrix Operations

Transpose[*m*]	transpose
Inverse[*m*]	matrix inverse
Det[*m*]	determinant
Minors[*m*]	matrix of minors
Minors[*m*, *k*]	k^{th} minors
Tr[*m*]	trace

Some basic matrix operations.

Transposing a matrix interchanges the rows and columns in the matrix. If you transpose an $m \times n$ matrix, you get an $n \times m$ matrix as the result.

Transposing a 2×3 matrix gives a 3×2 result.

In[1]:= **Transpose[{{a, b, c}, {ap, bp, cp}}]**

Out[1]= {{a, ap}, {b, bp}, {c, cp}}

Det[*m*] gives the determinant of a square matrix *m*. Minors[*m*] is the matrix whose $(i, j)^{th}$ element gives the determinant of the submatrix obtained by deleting the $(n - i + 1)^{th}$ row and the $(n - j + 1)^{th}$ column of *m*. The $(i, j)^{th}$ cofactor of *m* is $(-1)^{i+j}$ times the $(i, j)^{th}$ element of the matrix of minors.

Minors[*m*, *k*] gives the determinants of the $k \times k$ submatrices obtained by picking each possible set of *k* rows and *k* columns from *m*. Note that you can apply Minors to rectangular, as well as square, matrices.

Here is the determinant of a simple 2×2 matrix.

In[2]:= **Det[{{a, b}, {c, d}}]**

Out[2]= -b c + a d

This generates a 3×3 matrix, whose i, j^{th} entry is a[*i*, *j*].

In[3]:= **m = Array[a, {3, 3}]**

Out[3]= {{a[1, 1], a[1, 2], a[1, 3]},
         {a[2, 1], a[2, 2], a[2, 3]},
         {a[3, 1], a[3, 2], a[3, 3]}}

Here is the determinant of m.

In[4]:= **Det[m]**

Out[4]= -a[1, 3] a[2, 2] a[3, 1] + a[1, 2] a[2, 3] a[3, 1] +
         a[1, 3] a[2, 1] a[3, 2] - a[1, 1] a[2, 3] a[3, 2] -
         a[1, 2] a[2, 1] a[3, 3] + a[1, 1] a[2, 2] a[3, 3]

This gives a matrix of the minors of m.

```
In[5]:= Minors[m]
Out[5]= {{-a[1, 2] a[2, 1] + a[1, 1] a[2, 2],
          -a[1, 3] a[2, 1] + a[1, 1] a[2, 3],
          -a[1, 3] a[2, 2] + a[1, 2] a[2, 3]},
         {-a[1, 2] a[3, 1] + a[1, 1] a[3, 2],
          -a[1, 3] a[3, 1] + a[1, 1] a[3, 3],
          -a[1, 3] a[3, 2] + a[1, 2] a[3, 3]},
         {-a[2, 2] a[3, 1] + a[2, 1] a[3, 2],
          -a[2, 3] a[3, 1] + a[2, 1] a[3, 3],
          -a[2, 3] a[3, 2] + a[2, 2] a[3, 3]}}}
```

You can use `Det` to find the characteristic polynomial for a matrix. Section 3.7.9 of the complete *Mathematica Book* discusses ways to find eigenvalues and eigenvectors directly.

Here is a 3×3 matrix.

```
In[6]:= m = Table[ 1/(i + j), {i, 3}, {j, 3} ]
```

$$Out[6]= \left\{\left\{\frac{1}{2}, \frac{1}{3}, \frac{1}{4}\right\}, \left\{\frac{1}{3}, \frac{1}{4}, \frac{1}{5}\right\}, \left\{\frac{1}{4}, \frac{1}{5}, \frac{1}{6}\right\}\right\}$$

Following precisely the standard mathematical definition, this gives the characteristic polynomial for m.

```
In[7]:= Det[ m - x IdentityMatrix[3] ]
```

$$Out[7]= \frac{1}{43200} - \frac{131\,x}{3600} + \frac{11\,x^2}{12} - x^3$$

The *trace* or *spur* of a matrix `Tr[m]` is the sum of the terms on the leading diagonal.

This finds the trace of a simple 2×2 matrix.

```
In[8]:= Tr[{{a, b}, {c, d}}]
Out[8]= a + d
```

•

•

•

3.8 Numerical Operations on Data

■ 3.8.3 Fourier Transforms

A common operation in analyzing various kinds of data is to find the Fourier transform, or spectrum, of a list of values. The idea is typically to pick out components of the data with particular frequencies, or ranges of frequencies.

Fourier[{u_1, u_2, ... , u_n}]	Fourier transform
InverseFourier[{v_1, v_2, ... , v_n}]	
	inverse Fourier transform

Fourier transforms.

Here is some data, corresponding to a square pulse.

```
In[1]:= {-1, -1, -1, -1, 1, 1, 1, 1}

Out[1]= {-1, -1, -1, -1, 1, 1, 1, 1}
```

Here is the Fourier transform of the data. It involves complex numbers.

```
In[2]:= Fourier[%]

Out[2]= {0. + 0. i, -0.707107 - 1.70711 i,
         0. + 0. i, -0.707107 - 0.292893 i,
         0. + 0. i, -0.707107 + 0.292893 i, 0. + 0. i,
         -0.707107 + 1.70711 i}
```

Here is the inverse Fourier transform.

```
In[3]:= InverseFourier[%]

Out[3]= {-1., -1., -1., -1., 1., 1., 1., 1.}
```

Fourier works whether or not your list of data has a length which is a power of two.

```
In[4]:= Fourier[{1, -1, 1}]

Out[4]= {0.57735 + 0. i, 0.57735 - 1. i, 0.57735 + 1. i}
```

This generates a length-200 list containing a periodic signal with random noise added.

```
In[5]:= data = Table[ N[Sin[30 2 Pi n/200] + (Random[ ] - 1/2)],
           {n, 200} ] ;
```

The data looks fairly random if you plot it directly.

In[6]:= **ListPlot[data, PlotJoined -> True]**

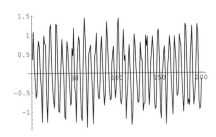

The Fourier transform, however, shows a strong peak at 30 + 1, and a symmetric peak at 201 − 30, reflecting the frequency component of the original signal near 30/200.

In[7]:= **ListPlot[Abs[Fourier[data]], PlotJoined -> True,**
                   PlotRange -> All]

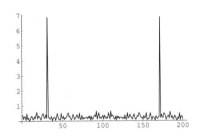

In *Mathematica*, the discrete Fourier transform v_s of a list u_r of length n is by default defined to be $\frac{1}{\sqrt{n}} \sum_{r=1}^{n} u_r e^{2\pi i(r-1)(s-1)/n}$. Notice that the zero frequency term appears at position 1 in the resulting list.

The inverse discrete Fourier transform u_r of a list v_s of length n is by default defined to be $\frac{1}{\sqrt{n}} \sum_{s=1}^{n} v_s e^{-2\pi i(r-1)(s-1)/n}$.

In different scientific and technical fields different conventions are often used for defining discrete Fourier transforms. The option `FourierParameters` in *Mathematica* allows you to choose any of these conventions you want.

common convention	setting	discrete Fourier transform	inverse discrete Fourier transform
Mathematica default	{0, 1}	$\frac{1}{n^{1/2}} \sum_{r=1}^{n} u_r e^{2\pi i(r-1)(s-1)/n}$	$\frac{1}{n^{1/2}} \sum_{s=1}^{n} v_s e^{-2\pi i(r-1)(s-1)/n}$
data analysis	{-1, 1}	$\frac{1}{n} \sum_{r=1}^{n} u_r e^{2\pi i(r-1)(s-1)/n}$	$\sum_{s=1}^{n} v_s e^{-2\pi i(r-1)(s-1)/n}$
signal processing	{1, -1}	$\sum_{r=1}^{n} u_r e^{-2\pi i(r-1)(s-1)/n}$	$\frac{1}{n} \sum_{s=1}^{n} v_s e^{2\pi i(r-1)(s-1)/n}$
general case	{a, b}	$\frac{1}{n^{(1-a)/2}} \sum_{r=1}^{n} u_r e^{2\pi ib(r-1)(s-1)/n}$	$\frac{1}{n^{(1+a)/2}} \sum_{s=1}^{n} v_s e^{-2\pi ib(r-1)(s-1)/n}$

Typical settings for `FourierParameters` with various conventions.

Fourier[{{u_{11}, u_{12}, ... }, {u_{21}, u_{22}, ... }, ... }]
                    two-dimensional Fourier transform

Two-dimensional Fourier transform.

Mathematica can find Fourier transforms for data in any number of dimensions. In n dimensions, the data is specified by a list nested n levels deep. Two-dimensional Fourier transforms are often used in image processing.

■ 3.8.4 Convolutions and Correlations

Convolution and correlation are central to many kinds of operations on lists of data. They are used in such areas as signal and image processing, statistical data analysis, approximations to partial differential equations, as well as operations on digit sequences and power series.

In both convolution and correlation the basic idea is to combine a kernel list with successive sublists of a list of data. The *convolution* of a kernel K_r with a list u_s has the general form $\sum_r K_r u_{s-r}$, while the *correlation* has the general form $\sum_r K_r u_{s+r}$.

ListConvolve[*kernel*, *list*]	form the convolution of *kernel* with *list*
ListCorrelate[*kernel*, *list*]	form the correlation of *kernel* with *list*

Convolution and correlation of lists.

This forms the convolution of the kernel {x, y} with a list of data.

```
In[1]:= ListConvolve[{x,y}, {a,b,c,d,e}]
Out[1]= {b x + a y, c x + b y, d x + c y, e x + d y}
```

This forms the correlation.

```
In[2]:= ListCorrelate[{x,y}, {a,b,c,d,e}]
Out[2]= {a x + b y, b x + c y, c x + d y, d x + e y}
```

In this case reversing the kernel gives exactly the same result as ListConvolve.

In[3]:= ListCorrelate[{y, x}, {a,b,c,d,e}]

Out[3]= {b x + a y, c x + b y, d x + c y, e x + d y}

This forms successive differences of the data.

In[4]:= ListCorrelate[{-1,1}, {a,b,c,d,e}]

Out[4]= {-a + b, -b + c, -c + d, -d + e}

In forming sublists to combine with a kernel, there is always an issue of what to do at the ends of the list of data. By default, ListConvolve and ListCorrelate never form sublists which would "overhang" the ends of the list of data. This means that the output you get is normally shorter than the original list of data.

With an input list of length 6, the output is in this case of length 4.

In[5]:= ListCorrelate[{1,1,1}, Range[6]]

Out[5]= {6, 9, 12, 15}

In practice one often wants to get output that is as long as the original list of data. To do this requires including sublists that overhang one or both ends of the list of data. The additional elements needed to form these sublists must be filled in with some kind of "padding". By default, *Mathematica* takes copies of the original list to provide the padding, thus effectively treating the list as being cyclic.

ListCorrelate[*kernel*, *list*]	do not allow overhangs on either side (result shorter than *list*)
ListCorrelate[*kernel*, *list*, 1]	allow an overhang on the right (result same length as *list*)
ListCorrelate[*kernel*, *list*, -1]	allow an overhang on the left (result same length as *list*)
ListCorrelate[*kernel*, *list*, {-1, 1}]	allow overhangs on both sides (result longer than *list*)
ListCorrelate[*kernel*, *list*, {k_L, k_R}]	allow particular overhangs on left and right

Controlling how the ends of the list of data are treated.

The default involves no overhangs.

In[6]:= ListCorrelate[{x, y}, {a, b, c, d}]

Out[6]= {a x + b y, b x + c y, c x + d y}

The last term in the last element now comes from the beginning of the list.

In[7]:= ListCorrelate[{x, y}, {a, b, c, d}, 1]

Out[7]= {a x + b y, b x + c y, c x + d y, d x + a y}

Now the first term of the first element and the last term of the last element both involve wraparound.

In[8]:= ListCorrelate[{x, y}, {a, b, c, d}, {-1, 1}]

Out[8]= {d x + a y, a x + b y, b x + c y, c x + d y, d x + a y}

In the general case ListCorrelate[*kernel*, *list*, {k_L, k_R}] is set up so that in the first element of the result, the first element of *list* appears multiplied by the element at position k_L in *kernel*, and in the last element of the result, the last element of *list* appears multiplied by the element at position

k_R in *kernel*. The default case in which no overhang is allowed on either side thus corresponds to ListCorrelate[*kernel*, *list*, {1, -1}].

With a kernel of length 3, alignments {-1, 2} always make the first and last elements of the result the same.

```
In[9]:= ListCorrelate[{x, y, z}, {a, b, c, d}, {-1, 2}]
Out[9]= {c x + d y + a z, d x + a y + b z, a x + b y + c z, b x + c y + d z,
         c x + d y + a z}
```

For many kinds of data, it is convenient to assume not that the data is cyclic, but rather that it is padded at either end by some fixed element, often 0, or by some sequence of elements.

⁺ ListCorrelate[*kernel*, *list*, *klist*, *p*]

                         pad with element *p*

⁺ ListCorrelate[*kernel*, *list*, *klist*, {p_1, p_2, ... }]

                         pad with cyclic repetitions of the p_i

⁺ ListCorrelate[*kernel*, *list*, *klist*, *list*]

                         pad with cyclic repetitions of the original data

⁺ ListCorrelate[*kernel*, *list*, *klist*, {}]

                         include no padding

Controlling the padding for a list of data.

This pads with element p.

```
In[10]:= ListCorrelate[{x, y}, {a, b, c, d}, {-1, 1}, p]
Out[10]= {p x + a y, a x + b y, b x + c y, c x + d y, d x + p y}
```

A common case is to pad with zero.

```
In[11]:= ListCorrelate[{x, y}, {a, b, c, d}, {-1, 1}, 0]
Out[11]= {a y, a x + b y, b x + c y, c x + d y, d x}
```

In this case q appears at one end, and p at the other.

```
In[12]:= ListCorrelate[{x, y}, {a, b, c, d}, {-1, 1}, {p, q}]
Out[12]= {q x + a y, a x + b y, b x + c y, c x + d y, d x + p y}
```

Different choices of kernel allow ListConvolve and ListCorrelate to be used for different kinds of computations.

This finds a moving average of data.

```
In[13]:= ListCorrelate[{1,1,1}/3, {a,b,c,d,e}, {-1,1}]
```

$$Out[13]= \left\{ \frac{a}{3} + \frac{d}{3} + \frac{e}{3}, \frac{a}{3} + \frac{b}{3} + \frac{e}{3}, \frac{a}{3} + \frac{b}{3} + \frac{c}{3}, \frac{b}{3} + \frac{c}{3} + \frac{d}{3}, \right.$$
$$\left. \frac{c}{3} + \frac{d}{3} + \frac{e}{3}, \frac{a}{3} + \frac{d}{3} + \frac{e}{3}, \frac{a}{3} + \frac{b}{3} + \frac{e}{3} \right\}$$

Here is a Gaussian kernel.

```
In[14]:= kern = Table[Exp[-n^2/100]/Sqrt[2. Pi], {n, -10, 10}] ;
```

This generates some "data".

```
In[15]:= data = Table[BesselJ[1, x] + 0.2 Random[ ], {x, 0, 10, .1}] ;
```

Here is a plot of the data.

In[16]:= **ListPlot[data]**

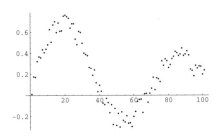

This convolves the kernel with the data.

In[17]:= **ListConvolve[kern, data, {-1, 1}] ;**

The result is a smoothed version of the data.

In[18]:= **ListPlot[%]**

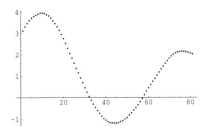

You can use **ListConvolve** and **ListCorrelate** to handle symbolic as well as numerical data.

This forms the convolution of two symbolic lists.

In[19]:= **ListConvolve[{a,b,c}, {u,v,w}, {-1, 1}, 0]**

Out[19]= {c u + b v + a w}

The result corresponds exactly with the coefficients in the expanded form of this product of polynomials.

In[20]:= **Expand[(a + b x + c x^2)(u + v x + w x^2)]**

Out[20]= $a u + b u x + a v x + c u x^2 + b v x^2 + a w x^2 + c v x^3 + b w x^3 + c w x^4$

ListConvolve and **ListCorrelate** work on data in any number of dimensions.

This imports image data from a file.

In[21]:= **g = ReadList["fish.data", Number, RecordLists->True];**

Here is the image.   *In[22]:=* **Show[Graphics[Raster[g], AspectRatio->Automatic]]**

This convolves the data with a   *In[23]:=* **ListConvolve[{{1,1,1},{1,-8,1},{1,1,1}}, g] ;**
two-dimensional kernel.

This shows the image corresponding to   *In[24]:=* **Show[Graphics[Raster[%], AspectRatio->Automatic]]**
the data.

RotateLeft[*list*, {d_1, d_2, ... }], RotateRight[*list*, {d_1, d_2, ... }]
              rotate elements cyclically by d_i positions at level i

⊹ PadLeft[*list*, {n_1, n_2, ... }], PadRight[*list*, {n_1, n_2, ... }]
              pad with zeros to create an $n_1 \times n_2 \times$... array

⊹ Take[*list*, m_1, m_2, ...], Drop[*list*, m_1, m_2, ...]
              take or drop m_i elements at level i

Other functions for manipulating multidimensional data.

A.2 Input Syntax

■ A.2.7 Operator Input Forms

Characters that are not letters, letter-like forms or structural elements are treated by *Mathematica* as *operators*. *Mathematica* has built-in rules for interpreting all operators. The functions to which these operators correspond may or may not, however, have built-in evaluation or other rules. Cases in which built-in meanings are by default defined are indicated by ◁ in the tables below.

Operators that construct two-dimensional boxes—all of which have names beginning with backslash—can only be used inside \(... \). The table below gives the interpretations of these operators within \!\(... \). Page 985 of the complete *Mathematica Book* gives interpretations when no \! is included.

expr and *expr$_i$*	any expression
symb	any symbol
patt	any pattern object
string and *string$_i$*	"*cccc*" or a sequence of letters, letter-like forms and digits
filename	like *string*, but can include additional characters described below
◁	built-in meanings exist
+	new in *Mathematica* Version 3.0
✢	new in *Mathematica* Version 4

Objects used in the tables of operator input forms.

operator form	full form	grouping
forms representing numbers (see page 969 of the complete Mathematica Book)		◁
forms representing symbols (see page 968 of the complete Mathematica Book)		◁
forms representing character strings (see page 968 of the complete Mathematica Book)		◁
expr::*string*	MessageName[*expr*, "*string*"]	◁
expr::*string$_1$*::*string$_2$*	MessageName[*expr*, "*string$_1$*", "*string$_2$*"]	◁

Operator input forms, in order of decreasing precedence, part one.

operator form	full form	grouping
forms containing # (*see page 978 of the complete Mathematica Book*)		◁
forms containing % (*see page 978 of the complete Mathematica Book*)		◁
forms containing _ (*see page 978 of the complete Mathematica Book*)		◁
<< *filename*	`Get["`*filename*`"]`	◁
+ $\overset{expr_2}{expr_1}$	`Overscript[`*expr₁*`, `*expr₂*`]`	$\overset{(\overset{e}{e})}{e}$
+ *expr₁* \& *expr₂*	`Overscript[`*expr₁*`, `*expr₂*`]`	*e*\&(*e*\&*e*)
+ $\underset{expr_2}{expr_1}$	`Underscript[`*expr₁*`, `*expr₂*`]`	$\underset{(\underset{e}{e})}{e}$
+ *expr₁* \+ *expr₂*	`Underscript[`*expr₁*`, `*expr₂*`]`	*e*\+(*e*\+*e*)
+ $\underset{expr_2}{\overset{expr_3}{expr_1}}$	`Underoverscript[`*expr₁*`, `*expr₂*`, `*expr₃*`]`	
+ *expr₁* \+ *expr₂* \% *expr₃*	`Underoverscript[`*expr₁*`, `*expr₂*`, `*expr₃*`]`	
+ *expr₁* \& *expr₂* \% *expr₃*	`Underoverscript[`*expr₁*`, `*expr₃*`, `*expr₂*`]`	
+ *expr₁* $_{expr_2}$	`Subscript[`*expr₁*`, `*expr₂*`]`	$e_{(e_e)}$
+ *expr₁* _ *expr₂*	`Subscript[`*expr₁*`, `*expr₂*`]`	*e*_(*e*_*e*)
+ *expr₁* _ *expr₂* \% *expr₃*	`Power[Subscript[`*expr₁*`, `*expr₂*`], `*expr₃*`]`	◁
+ \ ! *boxes*	(interpreted version of *boxes*)	
expr₁?*expr₂*	`PatternTest[`*expr₁*`, `*expr₂*`]`	◁
expr₁`[`*expr₂*`, ... ]`	*expr₁*`[`*expr₂*`, ... ]`	(*e*[*e*])[*e*]	◁
expr₁`[[`*expr₂*`, ... ]]`	`Part[`*expr₁*`, `*expr₂*`, ... ]`	(*e*[[*e*]])[[*e*]]	◁
+ *expr₁*⟦*expr₂*`, ... `⟧	`Part[`*expr₁*`, `*expr₂*`, ... ]`	(*e*⟦*e*⟧)⟦*e*⟧	◁
+ *expr₁* $_{⟦expr_2,...⟧}$	`Part[`*expr₁*`, `*expr₂*`, ... ]`	$(e_{⟦e⟧})_{⟦e⟧}$	◁
+ * *expr*	(boxes constructed from *expr*)	
expr++	`Increment[`*expr*`]`	◁
expr--	`Decrement[`*expr*`]`	◁
++*expr*	`PreIncrement[`*expr*`]`	◁
--*expr*	`PreDecrement[`*expr*`]`	◁
expr₁ @ *expr₂*	*expr₁*`[`*expr₂*`]`	*e* @ (*e* @ *e*)	◁

Operator input forms, in order of decreasing precedence, part two.

operator form	full form	grouping	
$+$ $expr_1$ $expr_2$ *(invisible application, input as $expr_1$ ⋮@⋮ $expr_2$)* $expr_1[expr_2]$			◁
$expr_1 \sim expr_2 \sim expr_3$	$expr_2[expr_1, expr_3]$	$(e \sim e \sim e) \sim e \sim e$	◁
$expr_1$ /@ $expr_2$	Map[$expr_1$, $expr_2$]	e /@ (e /@ e)	◁
$expr_1$ //@ $expr_2$	MapAll[$expr_1$, $expr_2$]	e //@ (e //@ e)	◁
$expr_1$ @@ $expr_2$	Apply[$expr_1$, $expr_2$]	e @@ (e @@ e)	◁
$+$ $expr_1$ @@@ $expr_2$	Apply[$expr_1$, $expr_2$, {1}]	e @@@ (e @@@ e)	◁
$expr!$	Factorial[$expr$]		◁
$expr!!$	Factorial2[$expr$]		◁
$expr'$	Derivative[1][$expr$]		◁
$expr_1$ <> $expr_2$ <> $expr_3$	StringJoin[$expr_1$, $expr_2$, $expr_3$]	e <> e <> e	◁
$expr_1$ ^ $expr_2$	Power[$expr_1$, $expr_2$]	$e\hat{}(e\hat{}e)$	◁
$+$ $expr_1{}^{expr_2}$	Power[$expr_1$, $expr_2$]	$e^{(e^e)}$	◁
$+$ $expr_1{}^{expr_3}_{expr_2}$	Power[Subscript[$expr_1$, $expr_2$], $expr_3$]		◁
$+$ $expr_1$ \^ $expr_2$ \% $expr_3$	Power[Subscript[$expr_1$, $expr_3$], $expr_2$]		◁
$+$ *vertical arrow and vector operators*			
$+$ $\sqrt{expr}$	Sqrt[$expr$]	$\sqrt{}(\sqrt{}e)$	◁
$+$ \@ $expr$	Sqrt[$expr$]	\@(\@e)	◁
$+$ \@ $expr$ \% n	Power[$expr$, 1/n]		◁
$+$ $\int expr_1 \, \mathrm{d}expr_2$	Integrate[$expr_1$, $expr_2$]	$\int (\int e \, \mathrm{d}\, e) \, \mathrm{d}\, e$	◁
$+$ $\int_{e_1}^{e_2} e_3 \, \mathrm{d}\, e_4$	Integrate[e_3, {e_4, e_1, e_2}]	$\int (\int e \, \mathrm{d}\, e) \, \mathrm{d}\, e$	◁
$+$ *other integration operators: see page 979 of the complete Mathematica Book*			
$+$ $\partial_{expr_1} expr_2$	D[$expr_2$, $expr_1$]	$\partial_e (\partial_e e)$	◁
$+$ $\nabla expr$	Del[$expr$]	$\nabla (\nabla e)$	
$+$ $\square\ expr$	Square[$expr$]	$\square (\square e)$	
$+$ $expr_1 \circ expr_2 \circ expr_3$	SmallCircle[$expr_1$, $expr_2$, $expr_3$]	$e \circ e \circ e$	
$+$ $expr_1 \odot expr_2 \odot expr_3$	CircleDot[$expr_1$, $expr_2$, $expr_3$]	$e \odot e \odot e$	

Operator input forms, in order of decreasing precedence, part three.

operator form	full form	grouping
+ $expr_1 \odot expr_2 \odot expr_3$	CircleDot[$expr_1$, $expr_2$, $expr_3$]	$e \odot e \odot e$
$expr_1$ ** $expr_2$ ** $expr_3$	NonCommutativeMultiply[$expr_1$, $expr_2$, $expr_3$]	e ** e ** e
+ $expr_1 \times expr_2 \times expr_3$	Cross[$expr_1$, $expr_2$, $expr_3$]	$e \times e \times e$ ◁
$expr_1 . expr_2 . expr_3$	Dot[$expr_1$, $expr_2$, $expr_3$]	$e . e . e$ ◁
$-expr$	Times[-1, $expr$]	◁
$+expr$	$expr$	◁
+ $\pm\, expr$	PlusMinus[$expr$]	
+ $\mp\, expr$	MinusPlus[$expr$]	
$expr_1 / expr_2$	$expr_1$ ($expr_2$)^-1	$(e / e) / e$ ◁
+ $expr_1 \div expr_2$	Divide[$expr_1$, $expr_2$]	$(e \div e) \div e$ ◁
+ $expr_1 \backslash/ expr_2$	Divide[$expr_1$, $expr_2$]	$(e\backslash/e)\backslash/e$ ◁
+ $expr_1 \backslash expr_2 \backslash expr_3$	Backslash[$expr_1$, $expr_2$, $expr_3$]	$e \backslash e \backslash e$
+ $expr_1 \diamond expr_2 \diamond expr_3$	Diamond[$expr_1$, $expr_2$, $expr_3$]	$e \diamond e \diamond e$
+ $expr_1 \wedge expr_2 \wedge expr_3$	Wedge[$expr_1$, $expr_2$, $expr_3$]	$e \wedge e \wedge e$
+ $expr_1 \vee expr_2 \vee expr_3$	Vee[$expr_1$, $expr_2$, $expr_3$]	$e \vee e \vee e$
+ $expr_1 \otimes expr_2 \otimes expr_3$	CircleTimes[$expr_1$, $expr_2$, $expr_3$]	$e \otimes e \otimes e$
+ $expr_1 \cdot expr_2 \cdot expr_3$	CenterDot[$expr_1$, $expr_2$, $expr_3$]	$e \cdot e \cdot e$
$expr_1\, expr_2\, expr_3$	Times[$expr_1$, $expr_2$, $expr_3$]	$e\, e\, e$ ◁
$expr_1 * expr_2 * expr_3$	Times[$expr_1$, $expr_2$, $expr_3$]	$e * e * e$ ◁
+ $expr_1 \times expr_2 \times expr_3$	Times[$expr_1$, $expr_2$, $expr_3$]	$e \times e \times e$ ◁
+ $expr_1 * expr_2 * expr_3$	Star[$expr_1$, $expr_2$, $expr_3$]	$e * e * e$
+ $\prod_{e_1=e_2}^{e_3} e_4$	Product[e_4, {e_1, e_2, e_3}]	$\prod (\prod e)$ ◁
+ $expr_1 \wr expr_2 \wr expr_3$	VerticalTilde[$expr_1$, $expr_2$, $expr_3$]	$e \wr e \wr e$
+ $expr_1 \sqcup expr_2 \sqcup expr_3$	Coproduct[$expr_1$, $expr_2$, $expr_3$]	$e \sqcup e \sqcup e$
+ $expr_1 \frown expr_2 \frown expr_3$	Cap[$expr_1$, $expr_2$, $expr_3$]	$e \frown e \frown e$
+ $expr_1 \smile expr_2 \smile expr_3$	Cup[$expr_1$, $expr_2$, $expr_3$]	$e \smile e \smile e$

Operator input forms, in order of decreasing precedence, part four.

operator form	full form	grouping	
$+$ $expr_1 \oplus expr_2 \oplus expr_3$	CirclePlus[$expr_1$, $expr_2$, $expr_3$]	$e \oplus e \oplus e$	
$+$ $expr_1 \ominus expr_2$	CircleMinus[$expr_1$, $expr_2$]	$(e \ominus e) \ominus e$	
$+$ $\sum_{e_1=e_2}^{e_3} e_4$	Sum[e_4, {e_1, e_2, e_3}]	$\sum (\sum e)$	◁
$expr_1 + expr_2 + expr_3$	Plus[$expr_1$, $expr_2$, $expr_3$]	$e + e + e$	◁
$expr_1 - expr_2$	$expr_1$ + (-1 $expr_2$)	$(e - e) - e$	◁
$+$ $expr_1 \pm expr_2$	PlusMinus[$expr_1$, $expr_2$]	$(e \pm e) \pm e$	
$+$ $expr_1 \mp expr_2$	MinusPlus[$expr_1$, $expr_2$]	$(e \mp e) \mp e$	
$+$ $expr_1 \cap expr_2$	Intersection[$expr_1$, $expr_2$]	$e \cap e \cap e$	◁
$+$ *other intersection operators*			
$+$ $expr_1 \cup expr_2$	Union[$expr_1$, $expr_2$]	$e \cup e \cup e$	◁
$+$ *other union operators*			
$expr_1$ == $expr_2$	Equal[$expr_1$, $expr_2$]	e == e == e	◁
$+$ $expr_1$ == $expr_2$	Equal[$expr_1$, $expr_2$]	e == e == e	◁
$expr_1$!= $expr_2$	Unequal[$expr_1$, $expr_2$]	e != e != e	◁
$+$ $expr_1 \neq expr_2$	Unequal[$expr_1$, $expr_2$]	$e \neq e \neq e$	◁
$+$ *other equality and similarity operators*			
$expr_1 > expr_2$	Greater[$expr_1$, $expr_2$]	$e > e > e$	◁
$expr_1$ >= $expr_2$	GreaterEqual[$expr_1$, $expr_2$]	e >= e >= e	◁
$+$ $expr_1 \geq expr_2$	GreaterEqual[$expr_1$, $expr_2$]	$e \geq e \geq e$	◁
$+$ $expr_1 \geqslant expr_2$	GreaterEqual[$expr_1$, $expr_2$]	$e \geqslant e \geqslant e$	◁
$expr_1 < expr_2$	Less[$expr_1$, $expr_2$]	$e < e < e$	◁
$expr_1$ <= $expr_2$	LessEqual[$expr_1$, $expr_2$]	e <= e <= e	◁
$+$ $expr_1 \leq expr_2$	LessEqual[$expr_1$, $expr_2$]	$e \leq e \leq e$	◁
$+$ $expr_1 \leqslant expr_2$	LessEqual[$expr_1$, $expr_2$]	$e \leqslant e \leqslant e$	◁
$+$ *other ordering operators*			
$+$ $expr_1$ ∣ $expr_2$	VerticalBar[$expr_1$, $expr_2$]	e ∣ e ∣ e	
$+$ $expr_1$ ∤ $expr_2$	NotVerticalBar[$expr_1$, $expr_2$]	e ∤ e ∤ e	
$+$ $expr_1$ ∥ $expr_2$	DoubleVerticalBar[$expr_1$, $expr_2$]	e ∥ e ∥ e	
$+$ $expr_1$ ∦ $expr_2$	NotDoubleVerticalBar[$expr_1$, $expr_2$]	e ∦ e ∦ e	
$+$ *horizontal arrow and vector operators*			
$+$ *diagonal arrow operators*			

Operator input forms, in order of decreasing precedence, part five.

operator form	full form	grouping	
$expr_1$ === $expr_2$	SameQ[$expr_1$, $expr_2$]	e === e === e	◁
$expr_1$ =!= $expr_2$	UnsameQ[$expr_1$, $expr_2$]	e =!= e =!= e	◁
+ $expr_1 \in expr_2$	Element[$expr_1$, $expr_2$]	$e \in e \in e$	◁
+ $expr_1 \subset expr_2$	Subset[$expr_1$, $expr_2$]	$e \subset e \subset e$	
+ $expr_1 \supset expr_2$	Superset[$expr_1$, $expr_2$]	$e \supset e \supset e$	
+ other set relation operators			
+ $\forall_{expr_1} expr_2$	ForAll[$expr_1$, $expr_2$]	$\forall_e$ ($\forall_e$ e)	
+ $\exists_{expr_1} expr_2$	Exists[$expr_1$, $expr_2$]	$\exists_e$ ($\exists_e$ e)	
+ $\nexists_{expr_1} expr_2$	NotExists[$expr_1$, $expr_2$]	$\nexists_e$ ($\nexists_e$ e)	
!$expr$	Not[$expr$]	!(!e)	◁
+ $\neg$ $expr$	Not[$expr$]	$\neg$ ($\neg$ e)	◁
$expr_1$ && $expr_2$ && $expr_3$	And[$expr_1$, $expr_2$, $expr_3$]	e && e && e	◁
+ $expr_1 \wedge expr_2 \wedge expr_3$	And[$expr_1$, $expr_2$, $expr_3$]	$e \wedge e \wedge e$	◁
$expr_1$ || $expr_2$ || $expr_3$	Or[$expr_1$, $expr_2$, $expr_3$]	e || e || e	◁
+ $expr_1 \vee expr_2 \vee expr_3$	Or[$expr_1$, $expr_2$, $expr_3$]	$e \vee e \vee e$	◁
+ $expr_1 \Rightarrow expr_2$	Implies[$expr_1$, $expr_2$]	$e \Rightarrow (e \Rightarrow e)$	◁
+ $expr_1 \Rightarrow expr_2$	Implies[$expr_1$, $expr_2$]	$e \Rightarrow (e \Rightarrow e)$	◁
+ $expr_1 \vdash expr_2$	RightTee[$expr_1$, $expr_2$]	$e \vdash (e \vdash e)$	
+ $expr_1 \vDash expr_2$	DoubleRightTee[$expr_1$, $expr_2$]	$e \vDash (e \vDash e)$	
+ $expr_1 \dashv expr_2$	LeftTee[$expr_1$, $expr_2$]	$(e \dashv e) \dashv e$	
+ $expr_1 \dashv expr_2$	DoubleLeftTee[$expr_1$, $expr_2$]	$(e \dashv e) \dashv e$	
+ $expr_1 \ni expr_2$	SuchThat[$expr_1$, $expr_2$]	$e \ni (e \ni e)$	
$expr$..	Repeated[$expr$]		◁
$expr$...	RepeatedNull[$expr$]		◁
$expr_1$ | $expr_2$	Alternatives[$expr_1$, $expr_2$]	e | e | e	◁
$symb$:$expr$	Pattern[$symb$, $expr$]		◁
$patt$:$expr$	Optional[$patt$, $expr$]		◁
$expr_1$ /; $expr_2$	Condition[$expr_1$, $expr_2$]	$(e/;e)/;e$	◁

Operator input forms, in order of decreasing precedence, part six.

operator form	full form	grouping	
$expr_1$ -> $expr_2$	`Rule[`$expr_1$`, `$expr_2$`]`	e -> (e -> e)	◁
+ $expr_1$ → $expr_2$	`Rule[`$expr_1$`, `$expr_2$`]`	e → (e → e)	◁
$expr_1$:> $expr_2$	`RuleDelayed[`$expr_1$`, `$expr_2$`]`	e :> (e :> e)	◁
+ $expr_1$:→ $expr_2$	`RuleDelayed[`$expr_1$`, `$expr_2$`]`	e :→ (e :→ e)	◁
$expr_1$ /. $expr_2$	`ReplaceAll[`$expr_1$`, `$expr_2$`]`	(e /. e) /. e	◁
$expr_1$ //. $expr_2$	`ReplaceRepeated[`$expr_1$`, `$expr_2$`]`	(e //. e) //. e	◁
$expr_1$ += $expr_2$	`AddTo[`$expr_1$`, `$expr_2$`]`		◁
$expr_1$ -= $expr_2$	`SubtractFrom[`$expr_1$`, `$expr_2$`]`		◁
$expr_1$ *= $expr_2$	`TimesBy[`$expr_1$`, `$expr_2$`]`		◁
$expr_1$ /= $expr_2$	`DivideBy[`$expr_1$`, `$expr_2$`]`		◁
$expr$ &	`Function[`$expr$`]`		◁
+ $expr_1$: $expr_2$	`Colon[`$expr_1$`, `$expr_2$`]`	e : e : e	
$expr_1$ // $expr_2$	$expr_2$`[`$expr_1$`]`	(e // e) // e	
+ $expr_1$ | $expr_2$	`VerticalSeparator[`$expr_1$`, `$expr_2$`]`	e | e | e	
+ $expr_1$ ∴ $expr_2$	`Therefore[`$expr_1$`, `$expr_2$`]`	e ∴ (e ∴ e)	
+ $expr_1$ ∵ $expr_2$	`Because[`$expr_1$`, `$expr_2$`]`	(e ∵ e) ∵ e	
$expr_1$ = $expr_2$	`Set[`$expr_1$`, `$expr_2$`]`	e = (e = e)	◁
$expr_1$:= $expr_2$	`SetDelayed[`$expr_1$`, `$expr_2$`]`		◁
$expr_1$ ^= $expr_2$	`UpSet[`$expr_1$`, `$expr_2$`]`		◁
$expr_1$ ^:= $expr_2$	`UpSetDelayed[`$expr_1$`, `$expr_2$`]`		◁
$symb$/: $expr_1$ = $expr_2$	`TagSet[`$symb$`, `$expr_1$`, `$expr_2$`]`		◁
$symb$/: $expr_1$:= $expr_2$	`TagSetDelayed[`$symb$`, `$expr_1$`, `$expr_2$`]`		◁
$expr$ =.	`Unset[`$expr$`]`		◁
$symb$/: $expr$ =.	`TagUnset[`$symb$`, `$expr$`]`		◁
$expr$ >> $filename$	`Put[`$expr$`, "`*filename*`"]`		◁
$expr$ >>> $filename$	`PutAppend[`$expr$`, "`*filename*`"]`		◁
$expr_1$;$expr_2$;$expr_3$	`CompoundExpression[`$expr_1$`, `$expr_2$`, `$expr_3$`]`		◁
$expr_1$;$expr_2$;	`CompoundExpression[`$expr_1$`, `$expr_2$`, Null]`		◁
+ $expr_1$ \` $expr_2$	`FormBox[`$expr_2$`, `$expr_1$`]`		

Operator input forms, in order of decreasing precedence, part seven.

A.3 Some General Notations and Conventions

■ A.3.5 Sequence Specifications

n	elements 1 through n
$-n$	last n elements
$\{n\}$	element n only
$\{m, n\}$	elements m through n (inclusive)
$\{m, n, s\}$	elements m through n in steps of s

Specifications for sequences of parts.

The sequence specification $\{m, n, s\}$ corresponds to elements $m, m + s, m + 2s, \ldots$, up to the largest element not greater than n.

Sequence specifications are used in the functions Drop, StringDrop, StringTake, Take and Thread.

A.9 Some Notes on Internal Implementation

◼ A.9.1 Introduction

General issues about the internal implementation of *Mathematica* are discussed on pages 215–223 of the complete *Mathematica Book*. Given here are brief notes on particular features.

~ These notes apply to Version 4. Algorithms and other aspects of implementation are subject to change in future versions.

It should be emphasized that these notes give only a rough indication of basic methods and algorithms used. The actual implementation usually involves many substantial additional elements.

Thus, for example, the notes simply say that DSolve solves second-order linear differential equations using the Kovacic algorithm. But the internal code which achieves this is over 60 pages long, includes a number of other algorithms, and involves a great many subtleties.

◼ A.9.2 Data Structures and Memory Management

A *Mathematica* expression internally consists of a contiguous array of pointers, the first to the head, and the rest to its successive elements.

Each expression contains a special form of hash code which is used both in pattern matching and evaluation.

For every symbol there is a central *symbol table entry* which stores all information about the symbol.

Most raw objects such as strings and numbers are allocated separately; unique copies are however maintained of small integers and of certain approximate numbers generated in computations.

Every piece of memory used by *Mathematica* maintains a count of how many times it is referenced. Memory is automatically freed when this count reaches zero.

The contiguous storage of elements in expressions reduces memory fragmentation and swapping. However, it can lead to the copying of a complete array of pointers when a single element in a long expression is modified. Many optimizations based on reference counts and pre-allocation are used to avoid such copying.

+ When appropriate, large lists and nested lists of numbers are automatically stored as packed arrays of machine-sized integers or real numbers. The *Mathematica* compiler is automatically used to compile complicated functions that will be repeatedly applied to such packed arrays. *MathLink*, DumpSave, Display, as well as various Import and Export formats, make external use of packed arrays.

◼ A.9.3 Basic System Features

Mathematica is fundamentally an interpreter which scans through expressions calling internal code pointed to by the symbol table entries of heads that it encounters.

Any transformation rule—whether given as *x* -> *y* or in a definition—is automatically compiled into a form which allows for rapid pattern matching. Many different types of patterns are distinguished and are handled by special code.

A form of hashing that takes account of blanks and other features of patterns is used in pattern matching.

The internal code associated with pattern matching is approximately 250 pages long.

When a large number of definitions are given for a particular symbol, a hash table is automatically built using a version of Dispatch so that appropriate rules can quickly be found.

■ A.9.4 Numerical and Related Functions

Number representation and numerical evaluation

~Large integers and high-precision approximate numbers are stored as arrays of base 2^{16} or 2^{32} digits, depending on the lengths of machine integers. ~■ Precision is internally maintained as a floating-point number. +■ IntegerDigits and related base conversion functions use a recursive divide-and-conquer algorithm. +■ N uses an adaptive procedure to increase its internal working precision in order to achieve whatever overall precision is requested. +■ Floor, Ceiling and related functions use an adaptive procedure similar to N to generate exact results from exact input.

Basic arithmetic

~Multiplication of large integers and high-precision approximate numbers is done using interleaved Karatsuba and FFT algorithms. ■ Integer powers are found by an algorithm based on Horner's rule. +■ Reciprocals and rational powers of approximate numbers use Newton's method. ■ Exact roots start from numerical estimates. ~■ Significance arithmetic is used for all arithmetic with approximate numbers beyond machine precision. ■ Basic arithmetic uses approximately 400 pages of C source code.

Pseudorandom numbers

Random uses the Wolfram rule 30 cellular automaton generator for integers. ■ It uses a Marsaglia-Zaman subtract-with-borrow generator for real numbers.

Number-theoretical functions

GCD uses the Jebelean-Weber accelerated GCD algorithm, together with a combination of Euclid's algorithm and an algorithm based on iterative removal of powers of 2. ■ PrimeQ first tests for divisibility using small primes, then uses the Miller-Rabin strong pseudoprime test base 2 and base 3, and then uses a Lucas test. ~■ As of 1997, this procedure is known to be correct only for $n < 10^{16}$, and it is conceivable that for larger n it could claim a composite number to be prime. ■ The package NumberTheory`PrimeQ` contains a much slower algorithm which has been proved correct for all n. It can return an explicit certificate of primality. ■ FactorInteger switches between removing small primes by trial division and using the Pollard $p - 1$, Pollard rho and continued fraction algorithm. ■ The package NumberTheory`FactorIntegerECM` contains an elliptic curve algorithm suitable for factoring some very large integers. ■ Prime and PrimePi use sparse caching and sieving. For large n, the Lagarias-Miller-Odlyzko algorithm for PrimePi is used, based on asymptotic estimates of the density of primes, and is inverted to give Prime. ■ LatticeReduce uses the Lenstra-Lenstra-Lovasz lattice reduction algorithm. +■ To find a requested number of terms ContinuedFraction uses a modification of Lehmer's indirect method, with a self-restarting divide-and-conquer algorithm to reduce the numerical precision required at each step. ~■ ContinuedFraction uses recurrence relations to find periodic continued fractions for quadratic irrationals. ~■ FromContinuedFraction uses iterated matrix multiplication optimized by a divide-and-conquer method.

Combinatorial functions

+Most combinatorial functions use sparse caching and recursion. ~■ Factorial, Binomial and related functions use a divide-and-conquer algorithm to balance the number of digits in subproducts. ■ Fibonacci[n] uses an iterative method based on the binary digit sequence of n. ■ PartitionsP[n] uses Euler's pentagonal formula for small n, and the non-recursive Hardy-Ramanujan-Rademacher method for larger n. ■ ClebschGordan and related functions use generalized hypergeometric series.

Elementary transcendental functions

Exponential and trigonometric functions use Taylor series, stable recursion by argument doubling, and functional relations. ■ Log and inverse trigonometric functions use Taylor series and functional relations.

Mathematical constants

Values of constants are cached once computed. +■ Binary splitting is used to subdivide computations of constants. ~■ Pi uses the Chudnovsky formula for computations up to ten million digits. ~■ E is computed from its series expansion. ■ EulerGamma uses the Brent-McMillan algorithm. ~■ Catalan is computed from a linearly convergent Ramanujan sum.

Special functions

For machine precision most special functions use *Mathematica*-derived rational minimax approximations. The notes that follow apply mainly to arbitrary precision. ■ Orthogonal polynomials use stable recursion formulas for polynomial cases and hypergeometric functions in general. ■ Gamma uses recursion, functional equations and the Binet asymptotic formula. ■ Incomplete gamma and beta functions use hypergeometric series and continued fractions. ■ PolyGamma uses Euler-Maclaurin summation, functional equations and recursion. ■ PolyLog uses Euler-Maclaurin summation, expansions in terms of incomplete gamma functions and numerical quadrature. ■ Zeta and related functions use Euler-Maclaurin summation and functional equations. Near the critical strip they also use the Riemann-Siegel formula. ■ StieltjesGamma uses Keiper's algorithm based on numerical quadrature of an integral representation of the zeta function. ■ The error function and functions related to exponential integrals are all evaluated using incomplete gamma functions. ■ The inverse error functions use binomial search and a high-order generalized Newton's method. ■ Bessel functions use series and asymptotic expansions. For integer orders, some also use stable forward recursion. ■ The hypergeometric functions use functional equations, stable recurrence relations, series expansions and asymptotic series. Methods from NSum and NIntegrate are also sometimes used. ■ ProductLog uses high-order Newton's method starting from rational approximations and asymptotic expansions. ■ Elliptic integrals are evaluated using the descending Gauss transformation. ■ Elliptic theta functions use series summation with recursive evaluation of series terms. ■ Other elliptic functions mostly use arithmetic-geometric mean methods. ■ Mathieu functions use Fourier series. The Mathieu characteristic functions use generalizations of Blanch's Newton method.

Numerical integration

With Method->Automatic, NIntegrate uses GaussKronrod in one dimension, and MultiDimensional otherwise. ■ If an explicit setting for MaxPoints is given, NIntegrate by default uses Method->QuasiMonteCarlo. ■ GaussKronrod: adaptive Gaussian quadrature with error estimation based on evaluation at Kronrod points. ■ DoubleExponential: non-adaptive double-exponential quadrature. ■ Trapezoidal: elementary trapezoidal method. ■ Oscillatory: transformation to handle integrals containing trigonometric and Bessel functions. ■ MultiDimensional: adaptive Genz-Malik algorithm. ■ MonteCarlo: non-adaptive Monte Carlo. ■ QuasiMonteCarlo: non-adaptive Halton-Hammersley-Wozniakowski algorithm.

Numerical sums and products

If the ratio test does not give 1, the Wynn epsilon algorithm is applied to a sequence of partial sums or products. ■ Otherwise Euler-Maclaurin summation is used with Integrate or NIntegrate.

Numerical differential equations

With Method->Automatic, NDSolve switches between a non-stiff Adams method and a stiff Gear method. Based on LSODE. ■ Adams: implicit Adams method with order between 1 and 12. ■ Gear: backward difference formula method with order between 1 and 5. ■ RungeKutta: Fehlberg order 4–5 Runge-Kutta method for non-stiff equations. ■ For linear boundary value problems the Gel'fand-Lokutsiyevskii chasing method is used. ■ For 1+1-dimensional PDEs the method of lines is used. ■ The code for NDSolve is about 500 pages long.

Approximate equation solving and minimization

Polynomial root finding is done based on the Jenkins-Traub algorithm. ■ For sparse linear systems, Solve and NSolve use several efficient numerical methods, mostly based on Gauss factoring with Markowitz products (approximately 250 pages of code). ~■ For systems of algebraic equations, NSolve computes a numerical Gröbner basis using an efficient monomial ordering, then uses eigenvalue methods to extract numerical roots. ■ FindRoot uses a damped Newton's method, the secant method and Brent's method. ■ With Method->Automatic, FindMinimum uses various methods due to Brent: the conjugate gradient in one dimension, and a modification of Powell's method in several dimensions. +■ If the function to be minimized is a sum of squares, FindMinimum uses the Levenberg-Marquardt method (Method->LevenbergMarquardt). ■ With Method->Newton FindMinimum uses Newton's method. With Method->QuasiNewton FindMinimum uses the BFGS version of the quasi-Newton method. ■ ConstrainedMax and related functions use an enhanced version of the simplex algorithm.

Data manipulation

~Fourier uses the FFT algorithm with decomposition of the length into prime factors. When the prime factors are large, fast convolution methods are used to maintain $O(n \log(n))$ asymptotic complexity. +■ For real input, Fourier uses a real transform method. +■ ListConvolve and ListCorrelate use FFT algorithms when possible. ■ InterpolatingFunction uses divided differences to construct Lagrange or Hermite interpolating polynomials. ■ Fit works by computing the product of the response vector with the pseudoinverse of the design matrix.

Approximate numerical linear algebra

Machine-precision matrices are typically converted to a special internal representation for processing. ■ Algorithms similar to those of LINPACK, EISPACK and LAPACK are used when appropriate. ■ LUDecomposition, Inverse, RowReduce and Det use Gaussian elimination with partial pivoting. LinearSolve uses the same methods, together with iterative improvement for high-precision numbers. ■ SingularValues uses the QR algorithm with Givens rotations. PseudoInverse and NullSpace are based on SingularValues. ■ QRDecomposition uses Householder transformations. ■ SchurDecomposition uses QR iteration. ■ MatrixExp uses Schur decomposition.

Exact numerical linear algebra

Inverse and LinearSolve use efficient row reduction based on numerical approximation. ■ With Modulus->n, modular Gaussian elimination is used. ■ Det uses modular methods and row reduction, constructing a result using the Chinese Remainder Theorem. ■ Eigenvalues works by interpolating the characteristic polynomial. ■ MatrixExp uses Putzer's method or Jordan decomposition.

■ A.9.5 Algebra and Calculus

Polynomial manipulation

For univariate polynomials, Factor uses a variant of the Cantor-Zassenhaus algorithm to factor modulo a prime, then uses Hensel lifting and recombination to build up factors over the integers. ■ Factoring over algebraic number fields is done by finding a primitive element over the rationals and then using Trager's algorithm. ■ For multivariate polynomials Factor works by substituting appropriate choices of integers for all but one variable, then factoring the resulting univariate polynomials, and reconstructing multivariate factors using Wang's algorithm. ■ The internal code for Factor exclusive of general polynomial manipulation is about 250 pages long. ■ FactorSquareFree works by finding a derivative and then iteratively computing GCDs. ■ Resultant uses either explicit subresultant polynomial remainder sequences or modular sequences accompanied by the Chinese Remainder Theorem. ■ Apart uses either a version of the Padé technique or the method of undetermined coefficients. ■ PolynomialGCD usually uses modular algorithms, including Zippel's sparse modular algorithm, but in some cases uses subresultant polynomial remainder sequences.

Symbolic linear algebra

RowReduce, LinearSolve and NullSpace are based on Gaussian elimination. ■ Inverse uses cofactor expansion and row reduction. Pivots are chosen heuristically by looking for simple expressions. ■ Det uses direct cofactor expansion for small matrices, and Gaussian elimination for larger ones. ■ MatrixExp finds eigenvalues and then uses Putzer's method. ■ Zero testing for various functions is done using symbolic transformations and interval-based numerical approximations after random numerical values have been substituted for variables.

Exact equation solving

For linear equations Gaussian elimination and other methods of linear algebra are used. ▪ Root objects representing algebraic numbers are usually isolated and manipulated using validated numerical methods. With ExactRootIsolation->True, Root uses for real roots a continued fraction version of an algorithm based on Descartes' rule of signs, and for complex roots the Collins-Krandick algorithm. ▪ For single polynomial equations, Solve uses explicit formulas up to degree four, attempts to reduce polynomials using Factor and Decompose, and recognizes cyclotomic and other special polynomials. ▪ For systems of polynomial equations, Solve constructs a Gröbner basis. ▪ Solve and GroebnerBasis use an efficient version of the Buchberger algorithm. ▪ For non-polynomial equations, Solve attempts to change variables and add polynomial side conditions. ▪ The code for Solve is about 500 pages long.

Simplification

~FullSimplify automatically applies about 40 types of general algebraic transformations, as well as about 400 types of rules for specific mathematical functions. ▪ Generalized hypergeometric functions are simplified using about 70 pages of *Mathematica* transformation rules. These functions are fundamental to many calculus operations in *Mathematica*. +▪ FunctionExpand uses an extension of Gauss's algorithm to expand trigonometric functions with arguments that are rational multiples of π. +▪ Simplify and FullSimplify cache results when appropriate. +▪ When assumptions specify that variables are real, methods based on cylindrical algebraic decomposition are used to deduce what transformations can be applied. +▪ For general polynomial inequalities, a version of the Collins algorithm is used with McCallum's improved projection operator. For strict inequalities, Strzebonski's method is used. For linear inequalities, methods based on either the simplex algorithm or the Loos-Weispfenning linear quantifier elimination algorithm are used. +▪ When assumptions involve equations among polynomials, Gröbner basis methods are used. +▪ For non-algebraic functions, a database of relations is used to determine the domains of function values from the domains of their arguments. Polynomial-oriented algorithms are used whenever the resulting domains correspond to semi-algebraic sets. +▪ For integer functions, several hundred theorems of number theory are used in the form of *Mathematica* rules.

Differentiation and integration

+Differentiation uses caching to avoid recomputing partial results. ▪ For indefinite integrals, an extended version of the Risch algorithm is used whenever both the integrand and integral can be expressed in terms of elementary functions, exponential integral functions, polylogarithms and other related functions. ▪ For other indefinite integrals, heuristic simplification followed by pattern matching is used. ▪ The algorithms in *Mathematica* cover all of the indefinite integrals in standard reference books such as Gradshteyn-Ryzhik. ▪ Definite integrals that involve no singularities are mostly done by taking limits of the indefinite integrals. ▪ Many other definite integrals are done using Marichev-Adamchik Mellin transform methods. The results are often initially expressed in terms of Meijer G functions, which are converted into hypergeometric functions using Slater's Theorem and then simplified. ▪ Integrate uses about 500 pages of *Mathematica* code and 600 pages of C code.

Differential equations

Linear equations with constant coefficients are solved using matrix exponentiation. ▪ Second-order linear equations with variable coefficients whose solutions can be expressed in terms of elementary functions and their integrals are solved using the Kovacic algorithm. ▪ Linear equations with polynomial coefficients are solved in terms of special functions by using Mellin transforms. ▪ When possible, nonlinear equations are solved by symmetry reduction techniques. For first-order equations classical techniques are used; for second-order equations and systems Bocharov techniques are used. ▪ For partial differential equations, separation of variables and symmetry reduction are used. ▪ DSolve uses about 300 pages of *Mathematica* code and 200 pages of C code.

Sums and products

Polynomial series are summed using Bernoulli and Euler polynomials. ▪ Series involving rational and factorial functions are summed using Adamchik techniques in terms of generalized hypergeometric functions, which are then simplified. ▪ Series involving polygamma functions are summed using integral representations. ▪ Dirichlet and related series are summed using pattern matching. ▪ For infinite series, d'Alembert and Raabe convergence tests are used. ▪ The algorithms in *Mathematica* cover at least 90% of the sums in standard reference books such as Gradshteyn-Ryzhik. ▪ Products are done primarily using pattern matching. ▪ Sum and Product use about 100 pages of *Mathematica* code.

Series and limits

Series works by recursively composing series expansions of functions with series expansions of their arguments. ▪ Limits are found from series and using other methods.

▪ A.9.6 Output and Interfacing

Graphics

Hidden-surface elimination for 3D graphics is done so as to be independent of display resolution. ▪ A custom-written PostScript interpreter is used to render graphics in the front end. ▪ Notebooks use a custom platform-independent bitmap image format.

Front end

The front end uses *MathLink* both for communication with the kernel, and for communication between its different internal components. ▪ All menu items and other functions in the front end are specified using *Mathematica* expressions. ▪ Configuration and preference files use *Mathematica* language format. ▪ The Help Browser is based on *Mathematica* notebooks generated from the same source code as this book.

Notebooks

Notebooks are represented as *Mathematica* expressions. ▪ Notebook files contain additional cached outline information in the form of *Mathematica* comments. This information makes possible efficient random access. ▪ Incremental saving of notebooks is done so as to minimize rewriting of data, moving data already written out whenever possible.
⁺▪ Platform-independent double-buffering is used by default to minimize flicker when window contents are updated.
⁺▪ Autoscrolling uses a control theoretical mechanism to optimize smoothness and controllability. ▪ All special characters are platform-independently represented using Unicode. Mapping tables are set up for specific Kanji and other fonts.
⁺▪ Spell checking and hyphenation are done using algorithms and a 100,000-word standard English dictionary, together with a 20,000-word technical dictionary, with 5000 *Mathematica* and other words added. Spelling correction is done using textual and phonetic metrics.

MathLink

In OSI terms, *MathLink* is a presentation-level protocol, which can be layered on top of any transport medium, both message-based and stream-based. ▪ *MathLink* encodes data in a compressed format when it determines that both ends of a link are on compatible computer systems. ▪ *MathLink* can transmit out-of-band data such as interrupts as well as *Mathematica* expressions. ▪ When possible *MathLink* is implemented using dynamically linked shared libraries.

Expression formatting

The front end uses a directed acyclic graph to represent the box structure of formatted expressions. ▪ Boxes are interpreted using a two-dimensional generalization of an operator precedence parser. ▪ Incremental parsing is used to minimize structure and display updating. ▪ Character spacing and positioning are determined from font data and operator tables. ▪ Line breaking is globally optimized throughout expressions, based on a method similar to the one used for text layout in TEX. ⁺▪ During input, line breaking is set up so that small changes to expressions rarely cause large-scale reformatting; if the input needs to jump, an elliptical cursor tracker momentarily appears to guide the eye. ▪ Expression formatting uses about 2000 pages of C code.

A.10 Listing of Major Built-in *Mathematica* Objects

■ Introduction

This section gives an alphabetical list of built-in objects which are supported in *Mathematica* Version 4.

The list does not include objects such as `CirclePlus` that are associated with operators such as ⊕, but which have no built-in values.

The list also does not include objects that are defined in *Mathematica* packages, even those distributed as a standard part of the *Mathematica* system.

Note also that options which appear only in a single built-in *Mathematica* function are sometimes not given as separate entries in the list.

A few objects in the list, mostly ones related to external operations, are not available on some computer systems.

+■	object or feature completely new after Version 2.0
~■	object or feature whose functionality was extensively changed after Version 2.0
+■	object or feature completely new in Version 4
~■	object or feature whose functionality was extensively changed for Version 4

New and modified objects and features in the listing.

Note that between Version 2.0 and Version 3.0, the internal code for a large fraction of all functions in *Mathematica* was modified in one way or another; between Version 3.0 and Version 4 a smaller but still substantial amount of internal code was changed. Even if an object is not indicated by +■, +■, ~■ or ~■ in this listing, therefore, it may well have changed in its efficiency or in the details of the results it gives.

This listing includes only standard built-in *Mathematica* objects that reside in the `System`` context. In a typical version of *Mathematica* there may be additional objects present both in the `System`` context, as well as in the `Developer`` and `Experimental`` contexts. For production work it is best to use only documented objects in the `System`` context, since the specifications of other objects may change in future versions. The online documentation for your version of *Mathematica* may contain information on `Developer`` and `Experimental`` objects. Further information is available at the Wolfram Research web site.

System`	built-in objects given in this listing
Developer`	advanced objects intended for *Mathematica* developers
Experimental`	objects provided on an experimental basis

Contexts for built-in objects.

In many versions of *Mathematica*, you can access the text given in this section directly, typically using the Help Browser (see page 56 of the complete *Mathematica Book*). Typing ?*F* to the *Mathematica* kernel will also give you the main description of the object *F* from this section.

More information on related packages mentioned in this listing can be found using the Help Browser, or by looking at *Standard Add-on Packages* published by Wolfram Research. Note that the specifications of functions in packages are subject to incompatible changes in future versions of *Mathematica*.

There are a total of 105 objects in this listing.

■ Conventions in This Listing

text in this style	literal *Mathematica* input that you type in as it is printed (e.g., function names)
text in this style	expressions that you fill in (e.g., function arguments)
object$_1$, *object*$_2$, ...	a sequence of any number of expressions
+	new since *Mathematica* Version 2.0
~	modified since *Mathematica* Version 2.0
+	new in *Mathematica* Version 4
~	modified in *Mathematica* Version 4

Conventions used in the list of built-in objects.

Note that for items modified in Version 4 this listing makes no distinction between those new in Version 3.0 and those not.

◾ AbsoluteOptions

`AbsoluteOptions[`*expr*`]` gives the absolute settings of options specified in an expression such as a graphics object.

`AbsoluteOptions[`*expr*`, `*name*`]` gives the absolute setting for the option *name*.

`AbsoluteOptions[`*expr*`, {`*name*$_1$`, `*name*$_2$`, ... }]` gives a list of the absolute settings for the options *name*$_i$.

`AbsoluteOptions[`*object*`]` gives the absolute settings for options associated with an external object such as a `NotebookObject`.

`AbsoluteOptions` gives the actual settings for options used internally by *Mathematica* when the setting given is `Automatic` or `All`. ▪ `AbsoluteOptions` returns lists of rules, just like `Options`. ▪ You can use `AbsoluteOptions` on graphics options such as `PlotRange` and `Ticks`. ▪ If you ask for `AbsoluteOptions[NotebookObject[... ], `*name*`]` the kernel will send a request to the front end to find the result. ▪ See also: `Options`, `FullGraphics`. ▪ Related package: `Utilities`FilterOptions`.

◾ Algebraics

`Algebraics` represents the domain of algebraic numbers, as in $x \in$ `Algebraics`.

Algebraic numbers are defined to be numbers that solve polynomial equations with rational coefficients. ▪ $x \in$ `Algebraics` evaluates immediately only for quantities x that are explicitly constructed from rational numbers, radicals and `Root` objects, or are known to be transcendental. ▪ `Simplify[`*expr* $\in$ `Algebraics]` can be used to try to determine whether an expression corresponds to an algebraic number. ▪ `Algebraics` is output in `TraditionalForm` as $\mathbb{A}$. ▪ See page 54 of this Addendum. ▪ See also: `Element`, `Simplify`, `Integers`, `Root`, `Extension`, `Reals`.

◾ All

`All` is a setting used for certain options.

In `Part` and related functions, `All` specifies all parts at a particular level.

For example, `PlotRange -> All` specifies that all points are to be included in a plot. ▪ See page 135 of the complete *Mathematica Book*. ▪ See also: `Automatic`, `None`, `Part`.

◾ AppellF1

`AppellF1[`a`, `b_1`, `b_2`, `c`, `x`, `y`]` is the Appell hypergeometric function of two variables $F_1(a; b_1, b_2; c; x, y)$.

Mathematical function (see Section A.3.10 of the complete *Mathematica Book*). ▪ $F_1(a; b_1, b_2; c; x, y)$ has series expansion $\sum_{m=0}^{\infty} \sum_{n=0}^{\infty} (a)_{m+n}(b_1)_m(b_2)_n/(m!n!(c)_{m+n})x^m y^n$. ▪ $F_1(a; b_1, b_2; c; x, y)$ reduces to $_2F_1(a, b; c; z)$ when $x = 0$ or $y = 0$. ▪ `AppellF1[`a`, `b_1`, `b_2`, `c`, `x`, `y`]` has singular lines in two-variable complex (x, y) space at $\mathrm{Re}(x) = 1$ and $\mathrm{Re}(y) = 1$, and has branch cut discontinuities along the rays from 1 to ∞ in x and y. ▪ `FullSimplify` and `FunctionExpand` include transformation rules for `AppellF1`. ▪ See page 48 of this Addendum. ▪ See also: `Hypergeometric2F1`.

■ Apply

Apply[*f*, *expr*] or *f* @@ *expr* replaces the head of *expr* by *f*.

Apply[*f*, *expr*, *levelspec*] replaces heads in parts of *expr* specified by *levelspec*.

Examples: Apply[f, {a, b, c}] ⟶ f[a, b, c]; Apply[Plus, g[a, b]] ⟶ a + b. ■ Level specifications are described on page 989 of the complete *Mathematica Book*. ■ The default value for *levelspec* in Apply is {0}. ⁺■ *f* @@@ *expr* is equivalent to Apply[*f*, *expr*, {1}]. ■ Examples: Apply[f, {{a,b},{c,d}}] ⟶ f[{a, b}, {c, d}]. ■ Apply[f, {{a,b},{c,d}}, {1}] ⟶ {f[a, b], f[c, d]}. ■ Apply[f, {{{{a}}}}, -2] ⟶ {f[f[f[a]]]}. ■ See page 27 of this Addendum and page 241 of the complete *Mathematica Book*. ■ See also: Map, Scan, Level, Operate, MapThread.

■ Array

Array[*f*, *n*] generates a list of length *n*, with elements *f*[*i*].

Array[*f*, {n_1, n_2, ... }] generates an $n_1 \times n_2 \times$ ... array of nested lists, with elements *f*[i_1, i_2, ...].

~ Array[*f*, {n_1, n_2, ... }, {r_1, r_2, ... }] generates a list using the index origins r_i (default 1).

Array[*f*, *dims*, *origin*, *h*] uses head *h*, rather than List, for each level of the array.

Examples: Array[f, 3] ⟶ {f[1], f[2], f[3]}. ■ Array[f, {2, 3}] ⟶ {{f[1, 1], f[1, 2], f[1, 3]}, {f[2, 1], f[2, 2], f[2, 3]}} generates a 2 × 3 matrix. ■ Array[#1^#2 &, {2, 2}] ⟶ {{1, 1}, {2, 4}}. ■ Array[f, 3, 0] ⟶ {f[0], f[1], f[2]} generates an array with index origin 0. ■ Array[f, 3, 1, Plus] ⟶ f[1] + f[2] + f[3]. ■ Note that the dimensions given to Array are *not* in standard *Mathematica* iterator notation. ■ See page 247 of the complete *Mathematica Book*. ■ See also: Table.

■ AutoIndent

AutoIndent is an option for Cell which specifies what automatic indentation should be done at the beginning of a new line after an explicit return character has been entered.

~■ Possible settings for AutoIndent are:

False	do no indentation
True	indent the same as the previous line
Automatic	indent according to the structure of the expression (default)

■ With AutoIndent->True, tabs or spaces used for indentation on the previous line are explicitly inserted at the beginning of the new line. ~■ With AutoIndent->Automatic, line breaks are always indicated by an IndentingNewLine character even if they were originally entered using RET or \[NewLine]. ■ Indentation after an \[IndentingNewLine] is automatically redone every time an expression is displayed. ■ The amount of indentation after an IndentingNewLine is determined by the settings for the LineIndent and LineIndentMaxFraction options. ■ See page 34 of this Addendum and page 588 of the complete *Mathematica Book*. ■ See also: LineIndent, ParagraphIndent, ShowAutoStyles.

⁺■ BitAnd

BitAnd[n_1, n_2, ...] gives the bitwise AND of the integers n_i.

Integer mathematical function (see Section A.3.10 of the complete *Mathematica Book*). ■ BitAnd[n_1, n_2, ...] yields the integer whose binary bit representation has ones at positions where the binary bit representations of all of the n_i have ones. ■ For negative integers BitOr assumes a two's complement representation. ■ See page 43 of this Addendum. ■ See also: BitOr, BitXor, BitNot, And, IntegerDigits, DigitCount.

■ BitNot

BitNot[*n*] gives the bitwise NOT of the integer *n*.

Integer mathematical function (see Section A.3.10 of the complete *Mathematica Book*). ■ BitNot[*n*] turns ones into zeros and vice versa in the binary bit representation of *n*. ■ Integers are assumed to be represented in two's complement form, with an unlimited number of digits, so that BitNot[*n*] is simply equivalent to $-1 - n$. ■ See page 43 of this Addendum. ■ See also: BitAnd, BitOr, BitXor, Not.

■ BitOr

BitAnd[n_1, n_2, ...] gives the bitwise OR of the integers n_i.

Integer mathematical function (see Section A.3.10 of the complete *Mathematica Book*). ■ BitOr[n_1, n_2, ...] yields the integer whose binary bit representation has ones at positions where the binary bit representations of any of the n_i have ones. ■ For negative integers BitOr assumes a two's complement representation. ■ See page 43 of this Addendum. ■ See also: BitAnd, BitXor, BitNot, Or, IntegerDigits.

■ BitXor

BitXor[n_1, n_2, ...] gives the bitwise XOR of the integers n_i.

Integer mathematical function (see Section A.3.10 of the complete *Mathematica Book*). ■ BitXor[n_1, n_2, ...] yields the integer whose binary bit representation has ones at positions where an odd number of the binary bit representations of the n_i have ones. ■ For negative integers BitXor assumes a two's complement representation. ■ See page 43 of this Addendum. ■ See also: BitAnd, BitOr, BitNot, Xor, IntegerDigits.

■ Booleans

Booleans represents the domain of booleans, as in $x \in$ Booleans.

The domain of booleans is taken to consist of the symbols True and False. ■ $x \in$ Booleans evaluates immediately only if x is explicitly True or False. ■ Simplify[*expr* ∈ Booleans] can be used to try to determine whether an expression is boolean, with no undetermined variables. ■ Boolean is output in TraditionalForm as $\mathbb{B}$. ■ See page 54 of this Addendum. ■ See also: Element, Simplify, True, False, Integers.

■ CarmichaelLambda

CarmichaelLambda[*n*] gives the Carmichael function $\lambda(n)$, defined as the smallest integer *m* such that $k^m \equiv 1 \bmod n$ for all *k* relatively prime to *n*.

Integer mathematical function (see Section A.3.10 of the complete *Mathematica Book*). ■ CarmichaelLambda returns unevaluated if there is no integer *m* satisfying the necessary conditions. ■ See page 40 of this Addendum. ■ See also: MultiplicativeOrder, EulerPhi, RealDigits.

■ ColorFunction

ColorFunction is an option for various graphics functions which specifies a function to apply to z values to determine the color to use for a particular x, y region.

ColorFunction is an option for Plot3D, ListPlot3D, DensityPlot, ContourPlot, Raster and related functions. ■ With the default setting ColorFunctionScaling -> True, the arguments provided for the function specified by ColorFunction are always scaled to be in the range 0 to 1. ■ With ColorFunctionScaling -> False original unscaled values are used. ■ The function specified by ColorFunction must return a CMYKColor, GrayLevel, Hue or RGBColor directive. ■ ColorFunction -> Automatic yields a range of gray levels. ■ ColorFunction -> Hue yields a range of colors. ■ In three-dimensional graphics, ColorFunction is used only with the option setting Lighting -> False. ■ See page 478 of the complete *Mathematica Book*. ■ See also: ColorFunctionScaling, Lighting, ColorOutput.

■ ColorFunctionScaling

ColorFunctionScaling is an option for various graphics functions which specifies whether the values provided to a color function should be scaled to lie between 0 and 1.

The default setting for ColorFunctionScaling is True. ■ With ColorFunctionScaling -> False original unscaled values are fed to the color function. ■ See also: ColorFunction.

■ CompiledFunction

CompiledFunction[*args*, *argregs*, *nregs*, *instr*, *func*] represents compiled code for evaluating a compiled function.

args is a list giving a pattern for the type of each argument to the function. The types are specified as in Compile. ■ *argregs* is a list of the registers into which actual argument values should be placed to evaluate the compiled code. ■ *nregs* is a list of the numbers of logical, integer, real, complex and tensor registers required in evaluating the compiled code. ■ *instr* is a list of actual compiled code instructions. ■ *func* is a *Mathematica* pure function to be used if no result can be obtained from the compiled code for any reason. ■ Compile generates a CompiledFunction object which can be executed by applying it to appropriate arguments. ■ CompiledFunction objects that are constructed explicitly can also be executed. Basic consistency checks are done when such objects are first evaluated by *Mathematica*. ■ The code in a CompiledFunction object is based on an idealized register machine. ■ See page 356 of the complete *Mathematica Book*. ■ See also: InterpolatingFunction.

■ Complexes

Complexes represents the domain of complex numbers, as in $x \in$ Complexes.

$x \in$ Complexes evaluates immediately only if x is a numeric quantity. ■ Simplify[*expr* $\in$ Complexes] can be used to try to determine whether an expression corresponds to a complex number. ■ The domain of real numbers is taken to be a subset of the domain of complex numbers. ■ Complexes is output in TraditionalForm as $\mathbb{C}$. ■ See page 54 of this Addendum. ■ See also: Element, Simplify, NumberQ, NumericQ, Complex, Reals. ■ Related package: Algebra`Quaternions`.

◾ ContinuedFraction

ContinuedFraction[x, n] generates a list of the first n terms in the continued fraction representation of x.

ContinuedFraction[x] generates a list of all terms that can be obtained given the precision of x.

The continued fraction representation $\{a_1, a_2, a_3, \dots\}$ corresponds to the expression $a_1 + 1/(a_2 + 1/(a_3 + \dots))$. ▪ x can be either an exact or an inexact number. ▪ Example: ContinuedFraction[Pi, 4] ⟶ {3, 7, 15, 1}. ▪ For exact numbers, ContinuedFraction[x] can be used if x is rational, or is a quadratic irrational. ▪ For quadratic irrationals, ContinuedFraction[x] returns a result of the form $\{a_1, a_2, \dots, \{b_1, b_2, \dots\}\}$, corresponding to an infinite sequence of terms, starting with the a_i, and followed by cyclic repetitions of the b_i. ▪ Since the continued fraction representation for a rational number has only a limited number of terms, ContinuedFraction[x, n] may yield a list with less than n elements in this case. ▪ For terminating continued fractions, $\{\dots, k\}$ is always equivalent to $\{\dots, k-1, 1\}$; ContinuedFraction returns the first of these forms. ▪ FromContinuedFraction[*list*] reconstructs a number from the result of ContinuedFraction. ▪ See page 41 of this Addendum. ▪ Implementation notes: see page 82. ▪ See also: FromContinuedFraction, IntegerDigits, Rationalize, Khinchin, RealDigits.

◾ Cos

Cos[z] gives the cosine of z.

Mathematical function (see Section A.3.10 of the complete *Mathematica Book*). ▪ The argument of Cos is assumed to be in radians. (Multiply by Degree to convert from degrees.) ▪ Cos is automatically evaluated when its argument is a simple rational multiple of π; for more complicated rational multiples, FunctionExpand can sometimes be used. ▪ See page 731 of the complete *Mathematica Book*. ▪ See also: ArcCos, Sec, TrigToExp, TrigExpand.

◾ Degree

Degree gives the number of radians in one degree. It has a numerical value of $\frac{\pi}{180}$.

You can multiply by Degree to convert from degrees to radians. ▪ Example: 30 Degree represents $30°$. ▪ Degree can be entered in StandardForm and InputForm as °, ⦂deg⦂ or \[Degree]. ▪ Degree is printed in StandardForm as °. ▪ See page 45 of this Addendum and page 735 of the complete *Mathematica Book*.

◾ Derivative

f' represents the derivative of a function f of one argument.

Derivative[n_1, n_2, $\dots$][f] is the general form, representing a function obtained from f by differentiating n_1 times with respect to the first argument, n_2 times with respect to the second argument, and so on.

f' is equivalent to Derivative[1][f]. ▪ f'' evaluates to Derivative[2][f]. ▪ You can think of Derivative as a *functional operator* which acts on functions to give derivative functions. ▪ Derivative is generated when you apply D to functions whose derivatives *Mathematica* does not know. ▪ *Mathematica* attempts to convert Derivative[n][f] and so on to pure functions. Whenever Derivative[n][f] is generated, *Mathematica* rewrites it as D[f[#]&, {#, n}]. If *Mathematica* finds an explicit value for this derivative, it returns this value. Otherwise, it returns the original Derivative form. ▪ Example: Cos' ⟶ -Sin[#1]& ▪ Derivative[$-n$][f] represents the nth indefinite integral of f. ▪ Derivative[$\{n_1, n_2, \dots\}$][f] represents the derivative of $f[\{x_1, x_2, \dots\}]$ taken n_i times with respect to x_i. In general, arguments given in lists in f can be handled by using a corresponding list structure in Derivative. ▪ N[$f'[x]$] will give a numerical approximation to a derivative. ▪ See page 807 of the complete *Mathematica Book*. ▪ See also: D, Dt.

⁺■ DigitCount

DigitCount[n, b, d] gives the number of d digits in the base b representation of n.

DigitCount[n, b] gives a list of the numbers of 1, 2, ..., $b - 1$, 0 digits in the base b representation of n.

DigitCount[n] gives a list of the numbers of 1, 2, ..., 9, 0 digits in the base 10 representation of n.

DigitCount[n] is equivalent to DigitCount[n, 10, Mod[Range[10],10]]. ■ Integer mathematical function (see Section A.3.10 of the complete *Mathematica Book*). ■ See page 42 of this Addendum. ■ See also: IntegerDigits, FromDigits, BitAnd, IntegerExponent.

⁺■ DiracDelta

DiracDelta[x] represents the Dirac delta function $\delta(x)$.

DiracDelta[x_1, x_2, ...] represents the multidimensional Dirac delta function $\delta(x_1, x_2, ...)$.

DiracDelta[x] returns 0 for all numeric x other than 0. ■ DiracDelta can be used in integrals, integral transforms and differential equations. ■ Some transformations are done automatically when DiracDelta appears in a product of terms. ■ DiracDelta[x_1, x_2, ...] returns 0 if any of the x_i are numeric and not 0. ■ DiracDelta has attribute Orderless. ■ For exact numeric quantities, DiracDelta internally uses numerical approximations to establish its result. This process can be affected by the setting of the global variable $MaxExtraPrecision. ■ See page 59 of this Addendum. ■ See also: UnitStep, If, PrincipalValue, Limit, KroneckerDelta.

⁺■ DiscreteDelta

DiscreteDelta[n_1, n_2, ...] gives the discrete delta function $\delta(n_1, n_2, ...)$, equal to 1 if all the n_i are zero, and 0 otherwise.

DiscreteDelta[0] gives 1; DiscreteDelta[n] gives 0 for other numeric n. ■ DiscreteDelta has attribute Orderless. ■ See page 62 of this Addendum. ■ See also: IdentityMatrix, UnitStep, If, Signature, DiracDelta.

⁻■ Drop

Drop[*list*, n] gives *list* with its first n elements dropped.

Drop[*list*, $-n$] gives *list* with its last n elements dropped.

Drop[*list*, {n}] gives *list* with its nth element dropped.

Drop[*list*, {m, n}] gives *list* with elements m through n dropped.

⁺ Drop[*list*, {m, n, s}] gives *list* with elements m through n in steps of s dropped.

⁺ Drop[*list*, seq_1, seq_2, ...] gives a nested list in which elements specified by seq_i have been dropped at level i in *list*.

Drop uses the standard *sequence specification* (see page 80). ■ Examples: Drop[{a,b,c,d,e}, 2] ⟶ {c, d, e}. ■ Drop[{a,b,c,d,e}, -3] ⟶ {a, b}. ⁺■ Drop[Range[7], {2, 5, 2}] ⟶ {1, 3, 5, 6, 7}. ■ Drop can be used on an object with any head, not necessarily List. ⁺■ Drop[*list*, seq_1, seq_2] effectively drops all elements except those in a submatrix of *list*. ⁺■ Example: Drop[{{a,b,c},{d,e,f}}, 1, -1] ⟶ {{d, e}}. ■ See page 7 of this Addendum and page 118 of the complete *Mathematica Book*. ■ See also: Rest, StringDrop, Take, Cases. ■ Related package: LinearAlgebra`MatrixManipulation`.

■ E

E is the exponential constant e (base of natural logarithms), with numerical value $\simeq 2.71828$.

Mathematical constant (see Section A.3.11 of the complete *Mathematica Book*). +■ E can be entered in StandardForm and InputForm as ⅇ, :ee: or \[ExponentialE]. +■ In StandardForm and TraditionalForm, E is printed as ⅇ. ■ See page 45 of this Addendum and page 735 of the complete *Mathematica Book*. ■ Implementation notes: see page 83. ■ See also: Exp.

■ Element

Element[x, *dom*] or $x \in dom$ asserts that x is an element of the domain *dom*.

Element[{x_1, x_2, ... }, *dom*] asserts that all the x_i are elements of *dom*.

Element[*patt*, *dom*] asserts that any expression matching the pattern *patt* is an element of *dom*.

$x \in dom$ can be entered as x :elem: *dom* or x \[Element] *dom*. ■ Element can be used to set up assumptions in Simplify and related functions. ■ Possible domains are:

Algebraics	algebraic numbers
Booleans	True or False
Complexes	complex numbers
Integers	integers
Primes	prime numbers
Rationals	rational numbers
Reals	real numbers

■ $x \in dom$ if possible evaluates immediately when x is numeric. ■ Examples: Pi $\in$ Algebraics $\longrightarrow$ False; Pi $\in$ Reals $\longrightarrow$ True. ■ (x_1 | x_2 | ...) $\in$ *dom* is equivalent to {x_1, x_2, ... } $\in$ *dom*. ■ {x_1, x_2, ... } $\in$ *dom* evaluates to (x_1 | x_2 | ...) $\in$ *dom* if its truth or falsity cannot immediately be determined. ■ See page 2 of this Addendum. ■ See also: Simplify, MemberQ, IntegerQ, Assumptions, Condition, PatternTest, Equal, Less.

⁺■ Export

Export["*file.ext*", *expr*] exports data to a file, converting it to a format corresponding to the file extension *ext*.

Export["*file*", *expr*, "*format*"] exports data to a file, converting it to the specified format.

Export can handle numerical and textual data, graphics, sounds, and material from notebooks. ■ The following basic formats are supported for numerical and textual data:

"Lines"	list of strings to be placed on separate lines
"List"	list of numbers or strings, to be placed on separate lines
"Table"	list of lists of numbers or strings, to be placed in a two-dimensional array (.dat)
"Text"	single string of ordinary characters (.txt)
"UnicodeText"	single string of 16-bit Unicode characters
"Words"	list of strings to be separated by spaces

■ In List and Table format, numbers are written in C or Fortran-like "E" notation when necessary.
■ Machine-precision numbers are by default written to six-digit precision. ■ In Table format, columns are separated by spaces. ■ Export["*name*.txt", *expr*] uses "Text" format. ■ Export["*name*.dat", *expr*] uses "Table" format.
■ The following additional formats are also supported for numerical and textual data:

"HDF"	Hierarchical Data Format (.hdf)
"MAT"	MAT matrix format (.mat)

■ All graphics formats in Export can handle any type of 2D or 3D *Mathematica* graphics. ■ They can also handle Notebook and Cell objects. ■ In some formats, lists of frames for animated graphics can be given. ■ The following options can be given when exporting graphics:

ImageResolution	Automatic	resolution in dpi for the image
ImageRotated	False	whether to rotate the image (landscape mode)
ImageSize	Automatic	absolute image size in printer's points

■ The following graphics formats are independent of the setting for ImageResolution:

"AI"	Adobe Illustrator format (.ai)
"EPS"	Encapsulated PostScript (.eps)
"MPS"	*Mathematica* abbreviated PostScript (.mps)
"PCL"	Hewlett-Packard printer control language (.pcl)
"PDF"	Adobe Acrobat portable document format (.pdf)
"PICT"	Macintosh PICT
"WMF"	Microsoft Windows metafile format (.wmf)

■ The following graphics formats depend on the setting for ImageResolution:

"BMP"	Microsoft bitmap format (.bmp)
"EPSI"	Encapsulated PostScript with image preview (.epsi)
"EPSTIFF"	Encapsulated PostScript with TIFF preview
"GIF"	GIF and animated GIF (.gif)
"JPEG"	JPEG (.jpg, .jpeg)
"MGF"	*Mathematica* system-independent raster graphics format
"PBM"	portable bitmap format (.pbm)
"PGM"	portable graymap format (.pgm)
"PNG"	PNG format (.png)
"PNM"	portable anymap format (.pnm)
"PPM"	portable pixmap format (.ppm)
"PSImage"	PostScript image format (.psi)
"TIFF"	TIFF (.tif, .tiff)
"XBitmap"	X window system bitmap (.xbm)

(continued)

+■ **Export** *(continued)*

- The following sound formats are supported:

"AIFF"   AIFF format (.aif, .aiff)

"AU" μ law encoding (.au)

"SND"   sound file format (.snd)

"WAV"   Microsoft wave format (.wav)

- The following general options can be given:

ByteOrdering	$ByteOrdering	what byte ordering to use for binary data
CharacterEncoding	Automatic	the encoding to use for text characters
ConversionOptions	{}	private options for specific formats

- Notebook and Cell objects, as well as any box expression obtained from ToBoxes, can be exported in the following formats:

"HTML"   HTML (.htm, .html)

"TeX"   TeX (.tex)

- These formats generate markup material which maintains much of the document structure that exists within *Mathematica*. ■ With these formats Export operates like HTMLSave and TeXSave. ■ Many details can be specified in the setting for ConversionOptions. ■ Possible formats accepted by Export are given in the list $ExportFormats. ■ Export["!*prog*", *expr*, "*format*"] exports data to a pipe. ■ See also: Import, ExportString, $ExportFormats, Display, Write, Put.

+■ **ExportString**

ExportString[*expr*, "*format*"] generates a string corresponding to *expr* exported in the specified format.

Many graphics, sound and binary formats yield strings containing non-printable characters. ■ See notes for Export. ■ See also: ImportString, DisplayString.

~■ **Fibonacci**

Fibonacci[*n*] gives the Fibonacci number F_n.

Fibonacci[*n*, *x*] gives the Fibonacci polynomial $F_n(x)$.

Integer mathematical function (see Section A.3.10 of the complete *Mathematica Book*). ■ The F_n satisfy the recurrence relation $F_n = F_{n-1} + F_{n-2}$ with $F_1 = F_2 = 1$. ■ For any complex value of n the F_n are given by the general formula $F_n = (\phi^n - (-\phi)^{-n})/\sqrt{5}$, where ϕ is the golden ratio. ■ The Fibonacci polynomial $F_n(x)$ is the coefficient of t^n in the expansion of $t/(1 - xt - t^2)$. ■ The Fibonacci polynomials satisfy the recurrence relation $F_n(x) = xF_{n-1}(x) + F_{n-2}(x)$. +■ FullSimplify and FunctionExpand include transformation rules for combinations of Fibonacci numbers with symbolic arguments when the arguments are specified to be integers using $n \in$ Integers. ■ See page 44 of this Addendum and page 727 of the complete *Mathematica Book*. ■ Implementation notes: see page 82. ■ See also: GoldenRatio.

▪ FileNames

FileNames[ ] lists all files in the current working directory.

FileNames["*form*"] lists all files in the current working directory whose names match the string pattern *form*.

FileNames[{"*form*$_1$", "*form*$_2$", ... }] lists all files whose names match any of the *form*$_i$.

FileNames[*forms*, {"*dir*$_1$", "*dir*$_2$", ... }] lists files with names matching *forms* in any of the directories *dir*$_i$.

FileNames[*forms*, *dirs*, *n*] includes files that are in subdirectories up to *n* levels down.

The string pattern "*form*" can contain the metacharacters specified on page 992 of the complete *Mathematica Book*. ▪ FileNames["*"] is equivalent to FileNames[ ]. ▪ FileNames[*forms*, *dirs*, Infinity] looks for files in all subdirectories of the *dirs*. ▪ The list of files returned by FileNames is sorted in the order generated by the function Sort. ▪ FileNames[*forms*, *dirs*, *n*] includes names of directories only if they appear exactly at level *n*. ~▪ The *forms* can include relative or absolute directory specifications, in addition to names of files. ▪ Setting the option IgnoreCase -> True makes FileNames treat lower- and upper-case letters in file names as equivalent. ▪ On operating systems such as MS-DOS, FileNames always treats lower- and upper-case letters in file names as equivalent. ▪ See page 613 of the complete *Mathematica Book*. ▪ See also: Directory, FileType, Get. ▪ Related package: Utilities`Package`.

▪ Fourier

Fourier[*list*] finds the discrete Fourier transform of a list of complex numbers.

~▪ The discrete Fourier transform v_s of a list u_r of length n is by default defined to be $\frac{1}{\sqrt{n}} \sum_{r=1}^{n} u_r e^{2\pi i(r-1)(s-1)/n}$.

▪ Note that the zero frequency term appears at position 1 in the resulting list. ~▪ Other definitions are used in some scientific and technical fields. +▪ Different choices of definitions can be specified using the option FourierParameters. +▪ With the setting FourierParameters -> {a, b} the discrete Fourier transform computed by Fourier is $\frac{1}{n^{(1-a)/2}} \sum_{r=1}^{n} u_r e^{2\pi i b(r-1)(s-1)/n}$. +▪ Some common choices for {a, b} are {0, 1} (default), {-1, 1} (data analysis), {1, -1} (signal processing). +▪ The setting $b = -1$ effectively corresponds to conjugating both input and output lists. +▪ To ensure a unique inverse discrete Fourier transform, |b| must be relatively prime to n. +▪ The list of data supplied to Fourier need not have a length equal to a power of two. ▪ The *list* given in Fourier[*list*] can be nested to represent an array of data in any number of dimensions. ▪ The array of data must be rectangular. ~▪ If the elements of *list* are exact numbers, Fourier begins by applying N to them. ▪ See page 66 of this Addendum and page 868 of the complete *Mathematica Book*. ▪ Implementation notes: see page 84. ▪ See also: InverseFourier, FourierTransform, Fit.

▪ FourierCosTransform

FourierCosTransform[*expr*, *t*, ω] gives the symbolic Fourier cosine transform of *expr*.

FourierCosTransform[*expr*, {t_1, t_2, ... }, {ω_1, ω_2, ... }] gives the multidimensional Fourier cosine transform of *expr*.

The Fourier cosine transform of a function $f(t)$ is by default defined to be $\sqrt{\frac{2}{\pi}} \int_0^\infty f(t) \cos(\omega t)\, dt$. ▪ Other definitions are used in some scientific and technical fields. ▪ Different choices of definitions can be specified using the option FourierParameters. ▪ With the setting FourierParameters->{a, b} the Fourier cosine transform computed by FourierCosTransform is $2\sqrt{\frac{|b|}{(2\pi)^{1-a}}} \int_0^\infty f(t) \cos(b\omega t)\, dt$. ▪ See notes for FourierTransform. ▪ See page 57 of this Addendum. ▪ See also: FourierSinTransform, FourierTransform, Fourier, InverseFourierCosTransform.

FourierSinTransform

FourierSinTransform[*expr*, *t*, ω] gives the symbolic Fourier sine transform of *expr*.

FourierSinTransform[*expr*, {t_1, t_2, ... }, {ω_1, ω_2, ... }] gives the multidimensional Fourier sine transform of *expr*.

The Fourier sine transform of a function $f(t)$ is by default defined to be $\sqrt{\frac{2}{\pi}} \int_0^\infty f(t) \sin(\omega t)\, dt$. ▪ Other definitions are used in some scientific and technical fields. ▪ Different choices of definitions can be specified using the option FourierParameters. ▪ With the setting FourierParameters->{a, b} the Fourier sine transform computed by FourierSinTransform is $2\sqrt{\frac{|b|}{(2\pi)^{1-a}}} \int_0^\infty f(t) \sin(b\omega t)\, dt$. ▪ See notes for FourierTransform. ▪ See page 57 of this Addendum. ▪ See also: FourierCosTransform, FourierTransform, Fourier, InverseFourierSinTransform.

FourierTransform

FourierTransform[*expr*, *t*, ω] gives the symbolic Fourier transform of *expr*.

FourierTransform[*expr*, {t_1, t_2, ... }, {ω_1, ω_2, ... }] gives the multidimensional Fourier transform of *expr*.

The Fourier transform of a function $f(t)$ is by default defined to be $\frac{1}{\sqrt{2\pi}} \int_{-\infty}^\infty f(t) e^{i\omega t}\, dt$. ▪ Other definitions are used in some scientific and technical fields. ▪ Different choices of definitions can be specified using the option FourierParameters. ▪ With the setting FourierParameters->{a, b} the Fourier transform computed by FourierTransform is $\sqrt{\frac{|b|}{(2\pi)^{1-a}}} \int_{-\infty}^\infty f(t) e^{ib\omega t}\, dt$. ▪ Some common choices for {a, b} are {0, 1} (default; modern physics), {1, -1} (pure mathematics; systems engineering), {1, 1} (classical physics), {0, -2 Pi} (signal processing). ▪ Assumptions and other options to Integrate can also be given in FourierTransform. ▪ FourierTransform[*expr*, *t*, ω] yields an expression depending on the continuous variable ω that represents the symbolic Fourier transform of *expr* with respect to the continuous variable *t*. Fourier[*list*] takes a finite list of numbers as input, and yields as output a list representing the discrete Fourier transform of the input. ▪ In TraditionalForm, FourierTransform is output using $\mathcal{F}$. ▪ See page 56 of this Addendum and page 93 of the complete *Mathematica Book*. ▪ See also: FourierSinTransform, FourierCosTransform, Fourier, InverseFourierTransform, LaplaceTransform, Integrate.

FromCharacterCode

FromCharacterCode[*n*] gives a string consisting of the character with integer code *n*.

FromCharacterCode[{n_1, n_2, ... }] gives a string consisting of the sequence of characters with codes n_i.

+ FromCharacterCode[{{n_{11}, n_{12}, ... }, {n_{21}, ... }, ... }] gives a list of strings.

+ FromCharacterCode[... , "*encoding*"] uses the specified character encoding.

~▪ The integer *n* must lie between 0 and 65535, as returned by ToCharacterCode. ▪ For *n* between 0 and 127, ToCharacterCode returns ASCII characters. ~▪ For *n* between 129 and 255, it returns ISO Latin-1 characters. ~▪ For other *n* it returns characters specified by the standard *Mathematica* encoding based on Unicode. ~▪ InputForm[FromCharacterCode[*n*]] gives the full name assigned to a special character with character code *n*. +▪ Whether a particular character generated by FromCharacterCode can be rendered on your output device will depend on what fonts and drivers you are using. +▪ Encodings supported in FromCharacterCode[... , "*encoding*"] are listed in the notes for $CharacterEncoding. ▪ See page 396 of the complete *Mathematica Book*. ▪ See also: ToCharacterCode, CharacterRange, $CharacterEncoding. ▪ Related package: Utilities`BinaryFiles`.

+■ **FromContinuedFraction**

FromContinuedFraction[*list*] reconstructs a number from the list of its continued fraction terms.

FromContinuedFraction[{a_1, a_2, a_3, ... }] returns $a_1 + 1/(a_2 + 1/(a_3 + ...))$. ■ The n_i can be symbolic. ■ FromContinuedFraction[{a_1, a_2, ... , {b_1, b_2, ... }}] returns the exact number whose continued fraction terms start with the a_i, then consist of cyclic repetitions of the b_i. ■ FromContinuedFraction acts as the inverse of ContinuedFraction. ■ See page 41 of this Addendum. ■ Implementation notes: see page 82. ■ See also: ContinuedFraction, Rationalize, FromDigits, Fold.

~■ **FullSimplify**

FullSimplify[*expr*] tries a wide range of transformations on *expr* involving elementary and special functions, and returns the simplest form it finds.

+ FullSimplify[*expr*, *assum*] does simplification using assumptions.

FullSimplify will always yield at least as simple a form as Simplify, but may take substantially longer. ~■ The following options can be given:

ComplexityFunction	Automatic	how to assess the complexity of each form generated
ExcludedForms	{ }	patterns specifying forms of subexpression that should not be touched
TimeConstraint	Infinity	for how many seconds to try doing any particular transformation
TransformationFunctions	Automatic	functions to try in transforming the expression

■ FullSimplify uses RootReduce on expressions that involve Root objects. ■ FullSimplify does transformations on most kinds of special functions. ■ See notes for Simplify. ■ See page 50 of this Addendum and pages 67 and 783 of the complete *Mathematica Book*. ■ Implementation notes: see page 85. ■ See also: Simplify, Factor, Expand, PowerExpand, ComplexExpand, TrigExpand, Element, FunctionExpand.

~■ **FunctionExpand**

FunctionExpand[*expr*] tries to expand out special and certain other functions in *expr*, when possible reducing compound arguments to simpler ones.

+ FunctionExpand[*expr*, *assum*] expands using assumptions.

FunctionExpand uses a large collection of rules. +■ FunctionExpand applies to certain trigonometric functions as well as special functions. ■ FunctionExpand is automatically called by FullSimplify. +■ Assumptions in FunctionExpand can be specified as in Simplify. +■ Example: FunctionExpand[*expr*, $x \in$ Reals] performs expansion assuming that x is real. ■ See page 52 of this Addendum and page 762 of the complete *Mathematica Book*. ■ See also: TrigExpand, TrigToExp, ComplexExpand, FullSimplify.

+■ **Glaisher**

Glaisher is Glaisher's constant with numerical value $\simeq 1.28243$.

Mathematical constant (see Section A.3.11 of the complete *Mathematica Book*). ■ Glaisher's constant A satisfies $\log(A) = \frac{1}{12} - \zeta'(-1)$, where ζ is the Riemann zeta function. ■ See page 45 of this Addendum. ■ See also: Zeta.

+ ■ HarmonicNumber

HarmonicNumber[n] gives the nth harmonic number H_n.

HarmonicNumber[n, r] gives the harmonic number $H_n^{(r)}$ of order r.

Mathematical function (see Section A.3.10 of the complete *Mathematica Book*). ■ The harmonic numbers are given by $H_n^{(r)} = \sum_1^n 1/i^r$ with $H_n = H_n^{(1)}$. ■ See page 44 of this Addendum. ■ See also: EulerGamma, PolyGamma, Zeta, Log.

~ ■ Hypergeometric2F1

Hypergeometric2F1[a, b, c, z] is the hypergeometric function $_2F_1(a, b; c; z)$.

Mathematical function (see Section A.3.10 of the complete *Mathematica Book*). ■ The $_2F_1$ function has the series expansion $_2F_1(a, b; c; z) = \sum_{k=0}^{\infty}(a)_k(b)_k/(c)_k \, z^k/k!$. ■ Hypergeometric2F1[a, b, c, z] has a branch cut discontinuity in the complex z plane running from 1 to ∞. + ■ FullSimplify and FunctionExpand include transformation rules for Hypergeometric2F1. ■ See page 48 of this Addendum and page 750 of the complete *Mathematica Book*. ■ See also: AppellF1, Hypergeometric1F1, HypergeometricPFQ, Hypergeometric2F1Regularized.

+ ■ Hyphenation

Hyphenation is an option for Cell which specifies whether to allow hyphenation for words of text.

The choice of hyphenation points is based when possible on dictionaries and algorithms for the language in which the text is specified. ■ See also: TextJustification, LanguageCategory.

~ ■ I

I represents the imaginary unit $\sqrt{-1}$.

Numbers containing I are converted to the type Complex. + ■ I can be entered in StandardForm and InputForm as i, ⌷ii⌷ or \[ImaginaryI]. + ■ j, ⌷jj⌷ and \[ImaginaryJ] can also be used. + ■ In StandardForm and TraditionalForm, I is output as i. ■ See page 45 of this Addendum and page 735 of the complete *Mathematica Book*. ■ See also: Re, Im, ComplexExpand, GaussianIntegers.

+ ■ Import

Import["*name*.*ext*"] imports data from a file, assuming that it is in the format indicated by the file extension *ext*, and converts it to a *Mathematica* expression.

Import["*file*", "*format*"] imports data in the specified format from a file.

Import attempts to give a *Mathematica* expression whose meaning is as close as possible to the data in the external file. ■ Import can handle textual and tabular data, as well as graphics and sounds. ■ The following basic formats are supported for textual and tabular data:

"Lines"	lines of text
"List"	lines consisting of numbers or strings
"Table"	two-dimensional array of numbers or strings
"Text"	string of ordinary characters
"UnicodeText"	string of 16-bit Unicode characters
"Words"	words separated by spaces or newlines

(continued)

⁺■ **Import** *(continued)*

■ "Text" and "UnicodeText" return single *Mathematica* strings. ■ "Lines" and "Words" return lists of *Mathematica* strings. ■ "List" returns a list of *Mathematica* numbers or strings. ■ "Table" return a list of lists of *Mathematica* numbers or strings. ■ In "List" and "Table" format, numbers can be read in C or Fortran-like "E" notation. ■ Numbers without explicit decimal points are returned as exact integers. ■ In "Table" format, columns can be separated by spaces or tabs. ■ In "Words" format, words can be separated by any form of white space. ■ Import["*name*.txt"] uses "Text" format. ■ Import["*name*.dat"] uses "Table" format. ■ The following additional formats are also supported for numerical data:

"HDF"   Hierarchical Data Format (.hdf)
"MAT"   MAT matrix format (.mat)

■ Two-dimensional graphics formats are imported as Graphics objects; sound formats are imported as Sound objects. ■ Animated graphics are imported as lists of Graphics objects. ■ The following formats yield expressions of the form Graphics[*data*, *opts*]:

"EPS"   Encapsulated PostScript (.eps)
"EPSI"   Encapsulated PostScript with image preview (.epsi)
"EPSTIFF"   Encapsulated PostScript with TIFF preview
"MPS"   *Mathematica* abbreviated PostScript (.mps)

■ The following formats yield expressions of the form Graphics[Raster[*data*], *opts*]:

"BMP"   Microsoft bitmap format (.bmp)
"GIF"   GIF and animated GIF (.gif)
"JPEG"   JPEG (.jpg, .jpeg)
"MGF"   *Mathematica* system-independent raster graphics format
"PBM"   portable bitmap format (.pbm)
"PGM"   portable graymap format (.pgm)
"PNG"   PNG format (.png)
"PNM"   portable anymap format (.pnm)
"PPM"   portable pixmap format (.ppm)
"PSImage"   PostScript image format (.psi)
"TIFF"   TIFF (.tif, .tiff)
"XBitmap"   X window system bitmap

■ Imported raster data normally consists of integers; ColorFunction is often used to specify a color map. ■ The following formats yield expressions of the form Sound[SampledSoundList[*data*, *r*]]:

"AIFF"   AIFF format (.aif, .aiff)
"AU" μ law encoding (.au)
"SND"   sound file format (.snd)
"WAV"   Microsoft wave format (.wav)

■ The following general options can be given:

ByteOrdering	$ByteOrdering	what byte ordering to use for binary data
CharacterEncoding	Automatic	the encoding to use for characters in text
ConversionOptions	{}	private options for specific formats
Path	$Path	the path to search for files

■ Possible formats accepted by Import are given in the list $ImportFormats. ■ Import["!*prog*", "*format*"] imports data from a pipe. ■ See also: Export, ImportString, $ImportFormats, ReadList.

⁺■ **ImportString**

ImportString["*data*", "*format*"] imports data in the specified format from a string.

See notes for Import. ■ See also: ExportString.

■ InputAliases

`InputAliases` is an option for cells and notebooks which specifies additional ⠿*name*⠿ aliases to be allowed on input.

The setting `InputAliases->{"`*name*₁`"->`*expr*₁`, ... }` specifies that the ⠿*name*ᵢ⠿ should be replaced on input by the corresponding *expr*ᵢ. ▪ The *expr*ᵢ should be strings or box expressions. ▪ See page 34 of this Addendum. ▪ See also: `InputAutoReplacements, $PreRead, Set`.

■ InputAutoReplacements

`InputAutoReplacements` is an option for cells and notebooks which specifies strings of characters that should be replaced immediately on input.

The default setting of `InputAutoReplacements` for Input styles typically includes such rules as `"->" -> "→"`. ▪ In expression input, automatic replacements can be performed only on strings of characters that correspond to complete input tokens. ▪ In textual input, automatic replacements can be performed on strings of alphanumeric characters delimited by spaces or other punctuation characters. ▪ When material is copied from a notebook to the clipboard, replacements specified by `ExportAutoReplacements` are by default performed. Typically these replacements include ones that reverse the action of the replacements in `InputAutoReplacements`. ▪ When material is pasted from the clipboard into a notebook, replacements specified by `ImportAutoReplacements` are by default performed. Typically these replacements are a subset of those given in `InputAutoReplacements`. ▪ See page 34 of this Addendum. ▪ See also: `InputAliases, $PreRead, Set`.

■ Integers

`Integers` represents the domain of integers, as in $x \in$ `Integers`.

$x \in$ `Integers` evaluates immediately only if x is a numeric quantity. ▪ `Simplify[`*expr* $\in$ `Integers]` can be used to try to determine whether an expression is an integer. ▪ `IntegerQ[`*expr*`]` tests only whether *expr* is manifestly an integer (i.e., has head `Integer`). ▪ `Integers` is output in `TraditionalForm` as $\mathbb{Z}$. ▪ See page 2 of this Addendum. ▪ See also: `Element, Simplify, IntegerQ, Reals, Primes, Algebraics`.

■ IntegerExponent

`IntegerExponent[`n`, `b`]` gives the highest power of b that divides n.

`IntegerExponent[`n`]` is equivalent to `IntegerExponent[`n`, 10]`. ▪ `IntegerExponent[`n`, `b`]` gives the number of trailing zeros in the digits of n in base b. ▪ See page 39 of this Addendum. ▪ See also: `IntegerDigits, FactorInteger, MantissaExponent, DigitCount, Exponent`.

■ InverseFourier

`InverseFourier[`*list*`]` finds the discrete inverse Fourier transform of a list of complex numbers.

The inverse Fourier transform u_r of a list v_s of length n is defined to be $\frac{1}{\sqrt{n}} \sum_{s=1}^{n} v_s e^{-2\pi i (r-1)(s-1)/n}$. ▪ Note that the zero frequency term must appear at position 1 in the input list. ▪ Other definitions are used in some scientific and technical fields. ▪ Different choices of definitions can be specified using the option `FourierParameters`. ▪ With the setting `FourierParameters -> {`a`, `b`}` the discrete Fourier transform computed by `Fourier` is $\frac{1}{n^{(1+a)/2}} \sum_{s=1}^{n} v_s e^{-2\pi i b(r-1)(s-1)/n}$. ▪ Some common choices for {a, b} are {0, 1} (default), {-1, 1} (data analysis), {1, -1} (signal processing). ▪ The setting $b = -1$ effectively corresponds to reversing both input and output lists. ▪ To ensure a unique discrete Fourier transform, |b| must be relatively prime to n. ▪ The list of data need not have a length equal to a power of two. ▪ The *list* given in `InverseFourier[`*list*`]` can be nested to represent an array of data in any number of dimensions. ▪ The array of data must be rectangular. +▪ If the elements of *list* are exact numbers, `InverseFourier` begins by applying `N` to them. ▪ See page 66 of this Addendum and page 868 of the complete *Mathematica Book*. ▪ See also: `Fourier, InverseFourierTransform`.

■ InverseFourierCosTransform

InverseFourierCosTransform[*expr*, ω, *t*] gives the symbolic inverse Fourier cosine transform of *expr*.

InverseFourierCosTransform[*expr*, {ω_1, ω_2, ... }, {t_1, t_2, ... }] gives the multidimensional inverse Fourier cosine transform of *expr*.

The inverse Fourier cosine transform of a function $F(\omega)$ is by default defined as $\sqrt{\frac{2}{\pi}} \int_0^\infty F(\omega)\cos(\omega t)\,d\omega$. ■ Other definitions are used in some scientific and technical fields. ■ Different choices of definitions can be specified using the option FourierParameters. ■ With the setting FourierParameters->{*a*, *b*} the inverse Fourier transform computed by InverseFourierCosTransform is $2\sqrt{\frac{|b|}{(2\pi)^{1+a}}} \int_0^\infty F(\omega)\cos(b\omega t)\,d\omega$. ■ See notes for InverseFourierTransform. ■ See page 57 of this Addendum. ■ See also: InverseFourierSinTransform, InverseFourierTransform, InverseFourier, InverseFourierTransform.

■ InverseFourierSinTransform

InverseFourierSinTransform[*expr*, ω, *t*] gives the symbolic inverse Fourier sine transform of *expr*.

InverseFourierSinTransform[*expr*, {ω_1, ω_2, ... }, {t_1, t_2, ... }] gives the multidimensional inverse Fourier sine transform of *expr*.

The inverse Fourier sine transform of a function $F(\omega)$ is by default defined as $\sqrt{\frac{2}{\pi}} \int_0^\infty F(\omega)\sin(\omega t)\,d\omega$. ■ Other definitions are used in some scientific and technical fields. ■ Different choices of definitions can be specified using the option FourierParameters. ■ With the setting FourierParameters->{*a*, *b*} the inverse Fourier transform computed by InverseFourierSinTransform is $2\sqrt{\frac{|b|}{(2\pi)^{1+a}}} \int_0^\infty F(\omega)\sin(b\omega t)\,d\omega$. ■ See notes for InverseFourierTransform. ■ See page 57 of this Addendum. ■ See also: InverseFourierCosTransform, InverseFourierTransform, InverseFourier, InverseFourierTransform.

■ InverseFourierTransform

InverseFourierTransform[*expr*, ω, *t*] gives the symbolic inverse Fourier transform of *expr*.

InverseFourierTransform[*expr*, {ω_1, ω_2, ... }, {t_1, t_2, ... }] gives the multidimensional inverse Fourier transform of *expr*.

The inverse Fourier transform of a function $F(\omega)$ is by default defined as $\frac{1}{\sqrt{2\pi}} \int_{-\infty}^\infty F(\omega)\,e^{-i\omega t}\,d\omega$. ■ Other definitions are used in some scientific and technical fields. ■ Different choices of definitions can be specified using the option FourierParameters. ■ With the setting FourierParameters->{*a*, *b*} the inverse Fourier transform computed by InverseFourierTransform is $\sqrt{\frac{|b|}{(2\pi)^{1+a}}} \int_{-\infty}^\infty F(\omega)\,e^{-ib\omega t}\,d\omega$. ■ Some common choices for {*a*, *b*} are {0, 1} (default; modern physics), {1, -1} (pure mathematics; systems engineering), {1, 1} (classical physics), {0, -2 Pi} (signal processing). ■ Assumptions and other options to Integrate can also be given in InverseFourierTransform. ■ InverseFourierTransform[*expr*, ω, *t*] yields an expression depending on the continuous variable *t* that represents the symbolic inverse Fourier transform of *expr* with respect to the continuous variable ω. InverseFourier[*list*] takes a finite list of numbers as input, and yields as output a list representing the discrete inverse Fourier transform of the input. ■ In TraditionalForm, InverseFourierTransform is output using $\mathcal{F}^{-1}$. ■ See page 56 of this Addendum and page 93 of the complete *Mathematica Book*. ■ See also: InverseFourierSinTransform, InverseFourierCosTransform, InverseFourier, FourierTransform, InverseLaplaceTransform, Integrate.

+■ **InverseLaplaceTransform**

InverseLaplaceTransform[*expr*, *s*, *t*] gives the inverse Laplace transform of *expr*.

InverseLaplaceTransform[*expr*, {s_1, s_2, ... }, {t_1, t_2, ... }] gives the multidimensional inverse Laplace transform of *expr*.

The inverse Laplace transform of a function $F(s)$ is defined to be $\frac{1}{2\pi i}\int_{\gamma-i\infty}^{\gamma+i\infty}F(s)e^{st}\,ds$, where γ is an arbitrary positive constant chosen so that the contour of integration lies to the right of all singularities in $F(s)$. ■ Assumptions and other options to Integrate can also be given in InverseLaplaceTransform. ■ In TraditionalForm, InverseLaplaceTransform is output using $\mathcal{L}^{-1}$. ■ See page 55 of this Addendum and page 92 of the complete *Mathematica Book*. ■ See also: LaplaceTransform, InverseFourierTransform, InverseZTransform, Integrate.

+■ **InverseZTransform**

InverseZTransform[*expr*, *z*, *n*] gives the inverse Z transform of *expr*.

The inverse Z transform of a function $F(z)$ is defined to be the contour integral $\frac{1}{2\pi i}\oint F(z)z^{n-1}\,dz$. ■ See page 58 of this Addendum. ■ See also: ZTransform, InverseLaplaceTransform.

+■ **Khinchin**

Khinchin is Khinchin's constant, with numerical value $\simeq 2.68545$.

Mathematical constant (see Section A.3.11 of the complete *Mathematica Book*). ■ Khinchin's constant (sometimes called Khintchine's constant) is given by $\prod_{s=1}^{\infty}(1+\frac{1}{s(s+2)})^{\log_2 s}$. ■ See page 45 of this Addendum. ■ See also: ContinuedFraction.

+■ **KroneckerDelta**

KroneckerDelta[n_1, n_2, ...] gives the Kronecker delta $\delta_{n_1 n_2 \ldots}$, equal to 1 if all the n_i are equal, and 0 otherwise.

KroneckerDelta[0] gives 1; KroneckerDelta[*n*] gives 0 for other numeric *n*. ■ KroneckerDelta has attribute Orderless. ■ See page 39 of this Addendum. ■ See also: IdentityMatrix, UnitStep, If, Signature, DiracDelta.

+■ **LanguageCategory**

LanguageCategory is an option for Cell which determines in what category of language the contents of the cell should be assumed to be for purposes of spell checking and hyphenation.

Possible settings for LanguageCategory are:

"Formula"	mathematical formula
"Mathematica"	*Mathematica* input
"NaturalLanguage"	human natural language
None	do no spell checking or hyphenation

■ LanguageCategory is normally set to "NaturalLanguage" for text cells, and to "Mathematica" for input and output cells. ■ LanguageCategory is more often set at the level of styles than at the level of individual cells. ■ See page 34 of this Addendum. ■ See also: $Language, Hyphenation, FormatType.

⁺■ LaplaceTransform

LaplaceTransform[*expr*, *t*, *s*] gives the Laplace transform of *expr*.

LaplaceTransform[*expr*, {t_1, t_2, ... }, {s_1, s_2, ... }] gives the multidimensional Laplace transform of *expr*.

The Laplace transform of a function $f(t)$ is defined to be $\int_0^\infty f(t)e^{-st}\,dt$. ■ The lower limit of the integral is effectively taken to be 0_-, so that the Laplace transform of the Dirac delta function $\delta(t)$ is equal to 1. ■ Assumptions and other options to Integrate can also be given in LaplaceTransform. ■ In TraditionalForm, LaplaceTransform is output using $\mathcal{L}$. ■ See page 55 of this Addendum and page 92 of the complete *Mathematica Book*. ■ See also: InverseLaplaceTransform, FourierTransform, ZTransform, Integrate.

⬛ ListConvolve

ListConvolve[*ker*, *list*] forms the convolution of the kernel *ker* with *list*.

ListConvolve[*ker*, *list*, *k*] forms the cyclic convolution in which the kth element of *ker* is aligned with each element in *list*.

ListConvolve[*ker*, *list*, {k_L, k_R}] forms the cyclic convolution whose first element contains *list*[[1]] *ker*[[k_L]] and whose last element contains *list*[[-1]] *ker*[[k_R]].

ListConvolve[*ker*, *list*, *klist*, *p*] forms the convolution in which *list* is padded at each end with repetitions of the element *p*.

ListConvolve[*ker*, *list*, *klist*, {p_1, p_2, ... }] forms the convolution in which *list* is padded at each end with cyclic repetitions of the p_i.

ListConvolve[*ker*, *list*, *klist*, *padding*, *g*, *h*] forms a generalized convolution in which *g* is used in place of Times and *h* in place of Plus.

ListConvolve[*ker*, *list*, *klist*, *padding*, *g*, *h*, *lev*] forms a convolution using elements at level *lev* in *ker* and *list*.

With kernel K_r and list a_s, ListConvolve[*ker*, *list*] computes $\sum_r K_r a_{s-r}$, where the limits of the sum are such that the kernel never overhangs either end of the list. ■ Example: ListConvolve[{x,y}, {a,b,c}] ⟶ {b x + a y, c x + b y}. ■ ListConvolve[*ker*, *list*] gives a result of length Length[*list*]-Length[*ker*]+1. ■ ListConvolve[*ker*, *list*] allows no overhangs and is equivalent to ListConvolve[*ker*, *list*, {-1, 1}]. ■ ListConvolve[*ker*, *list*, *k*] is equivalent to ListConvolve[*ker*, *list*, {*k*, *k*}]. ■ The values of k_L and k_R in ListConvolve[*ker*, *list*, {k_L, k_R}] determine the amount of overhang to allow at each end of *list*. ■ Common settings for {k_L, k_R} are:

{-1, 1}   no overhangs (default)
{-1, -1}   maximal overhang at the right-hand end
{1, 1}   maximal overhang at the left-hand end
{1, -1}   maximal overhangs at both beginning and end

■ Examples: ListConvolve[{x,y}, {a,b,c}, {1,1}] ⟶ {a x + c y, b x + a y, c x + b y}.
■ ListConvolve[{x,y}, {a,b,c}, {1,-1}] ⟶ {a x + c y, b x + a y, c x + b y, a x + c y}. ■ With maximal overhang at one end only, the result from ListConvolve is the same length as *list*.
■ ListConvolve[*ker*, *list*, {k_L, k_R}, *padlist*] effectively lays down repeated copies of *padlist*, then superimposes one copy of *list* on them and forms a convolution of the result. ■ Common settings for *padlist* are:

p	pad with repetitions of a single element
{p_1, p_2, ... }	pad with cyclic repetitions of a sequence of elements
list	pad by treating *list* as cyclic (default)
{}	do no padding

■ ListConvolve works with multidimensional kernels and lists of data.
■ ListConvolve[*ker*, *list*, {{k_{L1}, k_{L2}, ... }, {{k_{R1}, k_{R2}, ... }}]] forms the cyclic convolution whose {1,1,... } element contains *ker*[[k_{L1}, k_{L2}, ...]] *list*[[1,1,...]] and whose {-1,-1,... } element contains *ker*[[k_{R1}, k_{R2}, ...]] *list*[[-1,-1,...]]. ■ {k_L, k_R} is taken to be equivalent to {{k_L, k_L, ... }, {k_R, k_R, ... }}.
■ When a function *h* is specified to use in place of Plus, explicit nested *h* expressions are generated with a depth equal to the depth of *ker*. ■ ListConvolve works with exact numbers and symbolic data as well as approximate numbers. ■ Implementation notes: see page 84. ■ See page 68 of this Addendum. ■ See also: ListCorrelate, Partition, Inner, PadLeft.

■ ListCorrelate

ListCorrelate[*ker*, *list*] forms the correlation of the kernel *ker* with *list*.

ListCorrelate[*ker*, *list*, *k*] forms the cyclic correlation in which the kth element of *ker* is aligned with each element in *list*.

ListCorrelate[*ker*, *list*, {k_L, k_R}] forms the cyclic correlation whose first element contains *list*[[1]] *ker*[[k_L]] and whose last element contains *list*[[-1]] *ker*[[k_R]].

ListCorrelate[*ker*, *list*, *klist*, *p*] forms the correlation in which *list* is padded at each end with repetitions of the element *p*.

ListCorrelate[*ker*, *list*, *klist*, {p_1, p_2, ... }] forms the correlation in which *list* is padded at each end with cyclic repetitions of the p_i.

ListCorrelate[*ker*, *list*, *klist*, *padding*, *g*, *h*] forms a generalized correlation in which *g* is used in place of Times and *h* in place of Plus.

ListCorrelate[*ker*, *list*, *klist*, *padding*, *g*, *h*, *lev*] forms a correlation using elements at level *lev* in *ker* and *list*.

With kernel K_r and list a_s, ListCorrelate[*ker*, *list*] computes $\sum_r K_r a_{s+r}$, where the limits of the sum are such that the kernel never overhangs either end of the list. ■ Example: ListCorrelate[{x,y}, {a,b,c}] $\longrightarrow$ {a x + b y, b x + c y}. ■ For a one-dimensional list ListCorrelate[*ker*, *list*] is equivalent to ListConvolve[Reverse[*ker*], *list*]. ■ For higher-dimensional lists, *ker* must be reversed at every level. ■ See notes for ListConvolve. ■ Settings for k_L and k_R are negated in ListConvolve relative to ListCorrelate. ■ Common settings for {k_L, k_R} in ListCorrelate are:

{1, -1}	no overhangs (default)
{1, 1}	maximal overhang at the right-hand end
{-1, -1}	maximal overhang at the left-hand end
{-1, 1}	maximal overhangs at both beginning and end

■ See page 68 of this Addendum. ■ See also: ListConvolve, Partition, Inner, PadLeft.

■ MantissaExponent

MantissaExponent[*x*] gives a list containing the mantissa and exponent of a number *x*.

*MantissaExponent[*x*, *b*] gives the base-*b* mantissa and exponent of *x*.

Example: MantissaExponent[3.4 10^25] $\longrightarrow$ {0.34, 26}. ■ The mantissa always lies between $1/b$ and 1 or -1 and $-1/b$. ■ MantissaExponent works with exact as well as approximate numeric quantities. ■ Example: MantissaExponent[Exp[Pi], 2] $\longrightarrow \left\{ \frac{e^\pi}{32}, 5 \right\}$. ■ See page 702 of the complete *Mathematica Book*. ■ See also: Log, RealDigits, IntegerExponent.

~■ MessageName

symbol::*tag* is a name for a message.

You can specify messages by defining values for *symbol*::*tag*. ■ *symbol*::*tag* is converted to `MessageName[`*symbol*`, "`*tag*`"]`. *tag* can contain any characters that can appear in symbol names. *symbol*::"*tag*" can also be used. ■ Assignments for *s*::*tag* are stored in the `Messages` value of the symbol *s*. ~■ The following messages are typically defined for built-in functions:

f::`template`   a template showing a typical case of the function

f::`usage`   a description of how to use the function

■ ?*f* prints out the message *f*::`usage`. ■ When ?*form* finds more than one function, only the names of each function are printed. ■ You can switch on and off messages using `On[`*s*::*tag*`]` and `Off[`*s*::*tag*`]`.
■ `MessageName[`*symbol*`, "`*tag*`", "`*lang*`"]` or *symbol*::*tag*::*lang* represents a message in a particular language. ■ See page 458 of the complete *Mathematica Book*. ■ See also: `Message`, `MessageList`, `$MessageList`.

~■ Minors

+ `Minors[`*m*`]` gives the minors of a matrix *m*.

`Minors[`*m*`, `*k*`]` gives k^{th} minors.

+■ For an $n \times n$ matrix the $(i, j)^{\text{th}}$ element of `Minors[`*m*`]` gives the determinant of the matrix obtained by deleting the $(n - i + 1)^{\text{th}}$ row and the $(n - j + 1)^{\text{th}}$ column of *m*. ■ `Map[Reverse, Minors[`*m*`], {0,1}]` makes the $(i, j)^{\text{th}}$ element correspond to deleting the ith row and j^{th} column of *m*. ■ `Minors[`*m*`]` is equivalent to `Minors[`*m*`, `*n*`-1]`.
■ `Minors[`*m*`, `*k*`]` gives the determinants of the $k \times k$ submatrices obtained by picking each possible set of *k* rows and *k* columns from *m*. ~■ Each element in the result corresponds to taking rows and columns with particular lists of positions. The ordering of the elements is such that reading across or down the final matrix the successive lists of positions appear in lexicographic order. ■ For an $n_1 \times n_2$ matrix `Minors[`*m*`, `*k*`]` gives a $\binom{n_1}{k} \times \binom{n_2}{k}$ matrix.
+■ `Minors[`*m*`, `*k*`, `*f*`]` applies the function *f* rather than `Det` to each of the submatrices picked out. ■ See page 64 of this Addendum and page 844 of the complete *Mathematica Book*. ■ See also: `Det, Delete`.

~■ Mod

`Mod[`*m*`, `*n*`]` gives the remainder on division of *m* by *n*.

+ `Mod[`*m*`, `*n*`, `*d*`]` uses an offset *d*.

For integers *m* and *n* `Mod[`*m*`, `*n*`]` lies between 0 and $n - 1$. +■ `Mod[`*m*`, `*n*`, 1]` gives a result in the range 1 to *n*, suitable for use in functions such as `Part`. +■ `Mod[`*m*`, `*n*`, `*d*`]` gives a result *x* such that $d \le x < d + n$ and $x \bmod n = m \bmod n$. ■ The sign of `Mod[`*m*`, `*n*`]` is always the same as the sign of *n*, at least so long as *m* and *n* are both real. ■ `Mod[`*m*`, `*n*`]` is equivalent to $m - n$ `Quotient[`*m*`, `*n*`]`. +■ `Mod[`*m*`, `*n*`, `*d*`]` is equivalent to $m - n$ `Quotient[`*m*`, `*n*`, `*d*`]`. ~■ The arguments of `Mod` can be any numeric quantities, not necessarily integers.
~■ `Mod[`*x*`, 1]` gives the fractional part of *x*. ~■ For exact numeric quantities, `Mod` internally uses numerical approximations to establish its result. This process can be affected by the setting of the global variable `$MaxExtraPrecision`. ■ See page 39 of this Addendum and page 722 of the complete *Mathematica Book*. ■ See also: `PowerMod, Quotient, FractionalPart, MantissaExponent, PolynomialMod, PolynomialRemainder`.

◼ MultiplicativeOrder

MultiplicativeOrder[k, n] gives the multiplicative order of k modulo n, defined as the smallest integer m such that $k^m \equiv 1 \bmod n$.

MultiplicativeOrder[k, n, {r_1, r_2, ... }] gives the generalized multiplicative order of k modulo n, defined as the smallest integer m such that $k^m \equiv r_i \bmod n$ for any i.

Integer mathematical function (see Section A.3.10 of the complete *Mathematica Book*). ▪ MultiplicativeOrder returns unevaluated if there is no integer m satisfying the necessary conditions. ▪ See page 40 of this Addendum. ▪ See also: EulerPhi, PowerMod, CarmichaelLambda, RealDigits.

◼ NestWhile

NestWhile[f, *expr*, *test*] starts with *expr*, then repeatedly applies f until applying *test* to the result no longer yields True.

NestWhile[f, *expr*, *test*, m] supplies the most recent m results as arguments for *test* at each step.

NestWhile[f, *expr*, *test*, All] supplies all results so far as arguments for *test* at each step.

NestWhile[f, *expr*, *test*, m, *max*] applies f at most *max* times.

NestWhile[f, *expr*, *test*, m, *max*, n] applies f an extra n times.

NestWhile[f, *expr*, *test*, m, *max*, $-n$] returns the result found when f had been applied n fewer times.

NestWhile[f, *expr*, *test*] returns the first expression $f[f[... f[expr]...]]$ to which applying *test* does not yield True. ▪ If *test*[*expr*] does not yield True, NestWhile[f, *expr*, *test*] returns *expr*. ▪ NestWhile[f, *expr*, *test*, m] at each step evaluates *test*[res_1, res_2, ... , res_m]. It does not put the results res_i in a list. ▪ The res_i are given in the order they are generated, with the most recent coming last. ▪ NestWhile[f, *expr*, *test*, m] does not start applying *test* until at least m results have been generated. ▪ NestWhile[f, *expr*, *test*, {*mmin*, m}] does not start applying *test* until at least *mmin* results have been generated. At each step it then supplies as arguments to *test* as many recent results as possible, up to a maximum of m. ▪ NestWhile[f, *expr*, *test*, m] is equivalent to NestWhile[f, *expr*, *test*, {m, m}]. ▪ NestWhile[f, *expr*, UnsameQ, 2] is equivalent to FixedPoint[f, *expr*]. ▪ NestWhile[f, *expr*, *test*, All] is equivalent to NestWhile[f, *expr*, *test*, {1, Infinity}]. ▪ NestWhile[f, *expr*, UnsameQ, All] goes on applying f until the same result first appears more than once. ▪ NestWhile[f, *expr*, *test*, m, *max*, n] applies f an additional n times after *test* fails, or *max* applications have already been performed. ▪ NestWhile[f, *expr*, *test*, m, *max*, $-n$] is equivalent to Part[NestWhileList[f, *expr*, *test*, m, *max*], $-n$-1]. ▪ NestWhile[f, *expr*, *test*, m, Infinity, -1] returns, if possible, the last expression in the sequence *expr*, $f[expr]$, $f[f[expr]]$, ... for which *test* yields True. ▪ See page 26 of this Addendum. ▪ See also: NestWhileList, FixedPoint, Nest, While.

NestWhileList

`NestWhileList[`*f*`, `*expr*`, `*test*`]` generates a list of the results of applying *f* repeatedly, starting with *expr*, and continuing until applying *test* to the result no longer yields `True`.

`NestWhileList[`*f*`, `*expr*`, `*test*`, `*m*`]` supplies the most recent *m* results as arguments for *test* at each step.

`NestWhileList[`*f*`, `*expr*`, `*test*`, All]` supplies all results so far as arguments for *test* at each step.

`NestWhileList[`*f*`, `*expr*`, `*test*`, `*m*`, `*max*`]` applies *f* at most *max* times.

The last element of the list returned by `NestWhileList[`*f*`, `*expr*`, `*test*`]` is always an expression to which applying *test* does not yield True. ■ `NestWhileList[`*f*`, `*expr*`, `*test*`, `*m*`]` at each step evaluates *test*[res_1, res_2, ... , res_m]. It does not put the results res_i in a list. ■ The res_i are given in the order they are generated, with the most recent coming last. ■ `NestWhileList[`*f*`, `*expr*`, `*test*`, `*m*`]` does not start applying *test* until at least *m* results have been generated. ■ `NestWhileList[`*f*`, `*expr*`, `*test*`, {`*mmin*`, `*m*`}]` does not start applying *test* until at least *mmin* results have been generated. At each step it then supplies as arguments to *test* as many recent results as possible, up to a maximum of *m*. ■ `NestWhileList[`*f*`, `*expr*`, `*test*`, `*m*`]` is equivalent to `NestWhileList[`*f*`, `*expr*`, `*test*`, {`*m*`, `*m*`}]`. ■ `NestWhileList[`*f*`, `*expr*`, SameQ, 2]` is equivalent to `FixedPointList[`*f*`, `*expr*`]`. ■ `NestWhileList[`*f*`, `*expr*`, `*test*`, All]` is equivalent to `NestWhileList[`*f*`, `*expr*`, `*test*`, {1, Infinity}]`. ■ `NestWhileList[`*f*`, `*expr*`, UnsameQ, All]` goes on applying *f* until the same result first appears more than once. ■ `NestWhileList[`*f*`, `*expr*`, `*test*`, `*m*`, `*max*`, `*n*`]` applies *f* an extra *n* times, appending the results to the list generated. ■ `NestWhileList[`*f*`, `*expr*`, `*test*`, `*m*`, `*max*`, -`*n*`]` drops the last *n* elements from the list generated. ■ See page 26 of this Addendum. ■ See also: `NestWhile`, `FixedPointList`, `NestList`, `While`.

NotebookObject

`NotebookObject[`*fe*`, `*id*`]` is an object that represents an open notebook in the front end.

fe is a `FrontEndObject` which specifies the front end in which the notebook is open. ■ *id* is an integer that gives a unique serial number for this open notebook. ■ In `StandardForm` and `OutputForm` notebook objects are printed so as to indicate the current title of the window that would be used to display the notebook. ■ Functions such as `NotebookPrint` and `NotebookClose` take `NotebookObject` as their argument. ■ Within any open notebook, there is always a current selection. The current selection can be modified by applying functions such as `SelectionMove` to `NotebookObject`. ■ See page 554 of the complete *Mathematica Book*. ■ See also: `NotebookSelection`, `NotebookOpen`, `Notebooks`, `SelectedNotebook`, `NotebookTitle`.

■ PadLeft

PadLeft[*list*, *n*] makes a list of length *n* by padding *list* with zeros on the left.

PadLeft[*list*, *n*, *x*] pads by repeating the element *x*.

PadLeft[*list*, *n*, {x_1, x_2, ... }] pads by cyclically repeating the elements x_i.

PadLeft[*list*, *n*, *padding*, *m*] leaves a margin of *m* elements of padding on the right.

PadLeft[*list*, {n_1, n_2, ... }] makes a nested list with length n_i at level *i*.

PadLeft[*list*, *n*, ...] always returns a list of length *n*. ■ Example:
PadLeft[{a,b,c}, 7] ⟶ {0, 0, 0, 0, a, b, c}. ■ With padding {x_1, x_2, ... , x_s} cyclic repetitions of the x_i are effectively laid down and then the list is superimposed on top of them, with the last element of the list lying on an occurrence of x_s. ■ Examples: PadLeft[{a,b}, 7, {x,y,z}] ⟶ {z, x, y, z, x, a, b}.
■ PadLeft[{a,b}, 7, {x,y,z}, 2] ⟶ {y, z, x, a, b, x, y}. ■ PadLeft[*list*, *n*, *padding*, -*m*] truncates the last *m* elements of *list*. ■ A margin of Round[(*n*-Length[*list*])/2] effectively centers *list*. ■ PadLeft[*list*, *n*, *list*] effectively treats *list* as cyclic. ■ PadLeft[*list*, *n*, {*xlist*}] can be used to repeat an individual element that is itself a list. ■ Example: PadLeft[{a,b,c}, 5, {{u}}] ⟶ {{u}, {u}, a, b, c}. ■ PadLeft[{}, *n*, {x_1, x_2, ... }] repeats the sequence of x_i as many times as fits in a list of length *n*. ■ PadLeft[*list*, {n_1, n_2, ... }] creates a full array with dimensions {n_1, n_2, ... } even if *list* is ragged. ■ Negative n_i specify to pad on the right.
■ PadLeft[*list*, {n_1, n_2}, {{x_{11}, x_{12}, ... }, {x_{21}, ... }, ... }] pads by repeating the block of x_{ij}.
■ PadLeft[*list*, {n_1, n_2, ... }, *list*] effectively treats *list* as cyclic in every dimension.
■ PadLeft[*list*, {n_1, n_2, ... }, *padding*, {m_1, m_2, ... }] uses margin m_i at level *i*. ■ The object *list* need not have head List. ■ See page 11 of this Addendum. ■ See also: PadRight, Join, Partition, ListCorrelate, RotateLeft.

■ PadRight

PadRight[*list*, *n*] makes a list of length *n* by padding *list* with zeros on the right.

PadRight[*list*, *n*, *x*] pads by repeating the element *x*.

PadRight[*list*, *n*, {x_1, x_2, ... }] pads by cyclically repeating the elements x_i.

PadRight[*list*, *n*, *padding*, *m*] leaves a margin of *m* elements of padding on the left.

PadRight[*list*, {n_1, n_2, ... }] makes a nested list with length n_i at level *i*.

PadRight[*list*, *n*, ...] always returns a list of length *n*. ■ Example:
PadRight[{a,b,c}, 7] ⟶ {a, b, c, 0, 0, 0, 0}. ■ With padding {x_1, x_2, ... } cyclic repetitions of the x_i are effectively laid down and then the list is superimposed on top of them, with the first element of the list lying on an occurrence of x_1. ■ Examples: PadRight[{a,b}, 7, {x,y,z}] ⟶ {a, b, z, x, y, z, x}.
■ PadRight[{a,b}, 7, {x,y,z}, 2] ⟶ {y, z, a, b, z, x, y}. ■ See additional notes for PadLeft. ■ See page 11 of this Addendum. ■ See also: PadLeft, Join, Partition, ListCorrelate, RotateRight.

■ Part

expr[[*i*]] or Part[*expr*, *i*] gives the ith part of *expr*.

expr[[-*i*]] counts from the end.

expr[[0]] gives the head of *expr*.

expr[[*i*, *j*, ...]] or Part[*expr*, *i*, *j*, ...] is equivalent to *expr*[[*i*]] [[*j*]] ... .

expr[[{i_1, i_2, ... }]] gives a list of the parts i_1, i_2, ... of *expr*.

You can make an assignment like *t*[[*i*]] = *value* to modify part of an expression. ■ When *expr* is a list, *expr*[[{i_1, i_2, ... }]] gives a list of parts. In general, the head of *expr* is applied to the list of parts. ■ You can get a nested list of parts from *expr*[[*list*$_1$, *list*$_2$, ...]]. Each part has one index from each list. +■ If any of the *list*$_i$ are All, all parts at that level are kept. +■ *expr*[[All, *i*]] effectively gives the ith column in *expr*. ■ Notice that lists are used differently in Part than in functions like Extract, MapAt and Position. ■ *expr*[[Range[*i*, *j*]]] can be used to extract sequences of parts. +■ In StandardForm and InputForm, *expr*[[*spec*]] can be input as *expr*⟦*spec*⟧. +■ ⟦ and ⟧ can be entered as :[[: and :]]: or \[LeftDoubleBracket] and \[RightDoubleBracket]. +■ In StandardForm, *expr*[[*spec*]] can be input as *expr*$_{[[spec]]}$ or *expr*$_{⟦spec⟧}$ ■ See page 6 of this Addendum and page 233 of the complete *Mathematica Book*. ■ See also: First, Head, Last, Extract, Position, ReplacePart, MapAt, Take, PadLeft. ■ Related package: LinearAlgebra`MatrixManipulation`.

■ Partition

Partition[*list*, *n*] partitions *list* into non-overlapping sublists of length *n*.

Partition[*list*, *n*, *d*] generates sublists with offset *d*.

Partition[*list*, {n_1, n_2, ... }] partitions a nested list into blocks of size $n_1 \times n_2 \times$....

Partition[*list*, {n_1, n_2, ... }, {d_1, d_2, ... }] uses offset d_i at level *i* in *list*.

+ Partition[*list*, *n*, *d*, {k_L, k_R}] specifies that the first element of *list* should appear at position k_L in the first sublist, and the last element of *list* should appear at or after position k_R in the last sublist. If additional elements are needed, Partition fills them in by treating *list* as cyclic.

+ Partition[*list*, *n*, *d*, {k_L, k_R}, *x*] pads if necessary by repeating the element *x*.

+ Partition[*list*, *n*, *d*, {k_L, k_R}, {x_1, x_2, ... }] pads if necessary by cyclically repeating the elements x_i.

+ Partition[*list*, *n*, *d*, {k_L, k_R}, {}] uses no padding, and so can yield sublists of different lengths.

+ Partition[*list*, *nlist*, *dlist*, {*klist*$_L$, *klist*$_R$}, *padlist*] specifies alignments and padding in a nested list.

(continued)

~■ **Partition** *(continued)*

■ Example: Partition[{a,b,c,d,e,f}, 2] ⟶ {{a, b}, {c, d}, {e, f}}. ■ All the sublists generated by Partition[*list*, *n*, *d*] are of length *n*. Some elements at the end of *list* may therefore not appear in any sublist. ■ The element e in Partition[{a,b,c,d,e}, 2] ⟶ {{a, b}, {c, d}} is dropped.

■ Partition[{a,b,c,d,e}, 3, 1] ⟶ {{a, b, c}, {b, c, d}, {c, d, e}} generates sublists with offset 1. ■ All elements of *list* appear in the sublists generated by Partition[*list*, *n*, 1]. ■ If *d* is greater than *n* in Partition[*list*, *n*, *d*], then elements in the middle of *list* are skipped. ■ Partition[*list*, 1, *d*] picks out elements in the same way as Take[*list*, {1, -1, *d*}]. +■ Partition[*list*, *n*, *d*, {k_L, k_R}] effectively allows sublists that have overhangs that extend past the beginning or end of *list*. +■ Partition[*list*, *n*, *d*, *k*] is equivalent to Partition[*list*, *n*, *d*, {*k*, *k*}]. +■ Common settings for {k_L, k_R} are:

{1, -1}   allow no overhangs

{1, 1}   allow maximal overhang at the end

{-1, -1}   allow maximal overhang at the beginning

{-1, 1}   allow maximal overhangs at both beginning and end

+■ Example: Partition[{a,b,c,d},2,1,{-1,1}] ⟶ {{d, a}, {a, b}, {b, c}, {c, d}, {d, a}}.

+■ Partition[*list*, *n*, *d*, {k_L, k_R}, *padlist*] effectively lays down repeated copies of *padlist*, then superimposes one copy of *list* on them, and partitions the result. +■ Common settings for *padlist* are:

x   pad with repetitions of a single element

{x_1, x_2, ... }   pad with cyclic repetitions of a sequence of elements

list   pad by treating *list* as cyclic (default)

{}   do no padding, potentially leaving sublists of different lengths

+■ Example: Partition[{a,b,c,d},2,1,{-1,1},{x,y}] ⟶ {{y, a}, {a, b}, {b, c}, {c, d}, {d, x}}.

+■ Partition[{a,b,c,d},2,1,{-1,1},{}] ⟶ {{a}, {a, b}, {b, c}, {c, d}, {d}}. ■ If *list* has length *s*, then Partition[*list*, *n*, *d*] yields Max[0, Floor[(*s* + *d* - *n*)/*d*]] sublists. ■ Partition[*list*, {n_1, n_2, ... , n_r}] effectively replaces blocks of elements at level *r* in *list* by depth *r* nested lists of neighboring elements. ■ If no offsets are specified, the neighborhoods are adjacent and non-overlapping. ■ Partition[*list*, {n_1, n_2, ... }, *d*] uses offset *d* at every level. +■ Partition[*list*, *nlist*, *dlist*, {{k_{L1}, k_{L2}, ... }, {{k_{R1}, k_{R2}, ... }}}] specifies that element {1,1,... } of *list* should appear at position {k_{L1}, k_{L2}, ... } in the {1,1,... } block of the result, while element {-1,-1,... } of *list* should appear at or after position {k_{R1}, k_{R2}, ... } in the {-1,-1,... } block of the result. +■ {k_L, k_R} is taken to be equivalent to {{k_L, k_L, ... }, {k_R, k_R, ... }}. +■ {{k_1, k_2, ... }} is taken to be equivalent to {{k_1, k_2, ... }, {k_1, k_2, ... }}. +■ Partition[*list*, {n_1, n_2, ... , n_r}, *klist*, *padlist*] effectively makes a depth *r* array of copies of *padlist*, then superimposes *list* on them, and partitions the result. ■ If *list* has dimensions {s_1, s_2, ... , s_r} then Partition[*list*, {n_1, n_2, ... , n_r}] will have dimensions {q_1, q_2, ... , q_r, n_1, n_2, ... , n_r} where q_i is given by Floor[s_i/n_i]. ■ The object *list* need not have head List. ■ Partition[f[a,b,c,d], 2] ⟶ f[f[a, b], f[c, d]]. ■ See page 12 of this Addendum and page 124 of the complete *Mathematica Book*. ■ See also: Flatten, RotateLeft, Split, Take, PadLeft, ListConvolve.

+■ **Path**

Path is an option for Get and related functions which gives a list of directories to search in attempting to find an external file.

The default setting is Path :> $Path. ■ The possible settings for Path are the same as those for $Path. ■ See page 612 of the complete *Mathematica Book*. ■ See also: $Path, SetDirectory, $Input.

~■ PolyGamma

PolyGamma[z] gives the digamma function $\psi(z)$.

PolyGamma[n, z] gives the nth derivative of the digamma function $\psi^{(n)}(z)$.

PolyGamma[z] is the logarithmic derivative of the gamma function, given by $\psi(z) = \Gamma'(z)/\Gamma(z)$. ■ PolyGamma[$n$, z] is given by $\psi^{(n)}(z) = d^n\psi(z)/dz^n$. ■ The digamma function is $\psi(z) = \psi^{(0)}(z)$; $\psi^{(n)}(z)$ is the $(n+1)^{\text{th}}$ logarithmic derivative of the gamma function. ■ PolyGamma[z] and PolyGamma[n, z] are meromorphic functions of z with no branch cut discontinuities. +■ FullSimplify and FunctionExpand include transformation rules for PolyGamma. ■ See page 740 of the complete *Mathematica Book*. ■ Implementation notes: see page 83. ■ See also: Gamma, LogGamma, EulerGamma.

~■ PolyLog

PolyLog[n, z] gives the polylogarithm function $\mathrm{Li}_n(z)$.

+ PolyLog[n, p, z] gives the Nielsen generalized polylogarithm function $S_{n,p}(z)$.

Mathematical function (see Section A.3.10 of the complete *Mathematica Book*). ■ $\mathrm{Li}_n(z) = \sum_{k=1}^{\infty} z^k/k^n$.

■ $S_{n,p}(z) = (-1)^{n+p-1}/((n-1)!p!) \int_0^1 \log^{n-1}(t)\log^p(1-zt)/t\,dt$. +■ $S_{n-1,1}(z) = \mathrm{Li}_n(z)$. ■ PolyLog[$n$, z] has a branch cut discontinuity in the complex z plane running from 1 to ∞. +■ FullSimplify and FunctionExpand include transformation rules for PolyLog. ■ See page 46 of this Addendum and page 742 of the complete *Mathematica Book*. ■ Implementation notes: see page 83. ■ See also: Zeta, PolyGamma, LerchPhi.

+■ Primes

Primes represents the domain of primes numbers, as in $x \in$ Primes.

$x \in$ Primes evaluates only if x is a numeric quantity. ■ Simplify[*expr* $\in$ Primes] can be used to try to determine whether an expression corresponds to a prime number. ■ The domain of primes is taken to be a subset of the domain of integers. ■ PrimeQ[*expr*] returns False unless *expr* explicitly has head Integer. ■ Primes is output in TraditionalForm as $\mathbb{P}$. ■ See page 2 of this Addendum. ■ See also: Element, Simplify, PrimeQ, Prime, Integers.

~■ Quotient

Quotient[m, n] gives the integer quotient of m and n.

+ Quotient[m, n, d] uses an offset d.

Integer mathematical function (see Section A.3.10 of the complete *Mathematica Book*). ■ Quotient[m, n] is equivalent to Floor[m/n] for integers m and n. +■ Quotient[m, n, d] gives a result x such that $d \le m - nx < d + n$. ■ n*Quotient[m, n, d] + Mod[m, n, d] is always equal to m. ■ See page 39 of this Addendum and page 722 of the complete *Mathematica Book*. ■ See also: Mod, PolynomialQuotient.

~■ Raster

Raster[{{a_{11}, a_{12}, ... }, ... }] is a two-dimensional graphics primitive which represents a rectangular array of gray cells.

Raster[*array*, ColorFunction -> *f*] specifies that each cell should be rendered using the graphics directives obtained by applying the function *f* to the scaled value of the cell. ■ Raster[*array*, ColorFunction -> Hue] generates an array in which cell values are specified by hues. +■ With the option ColorFunctionScaling -> False the original cell values a_{ij}, rather than scaled cell values, are fed to the color function. ■ With the default setting ColorFunctionScaling -> True cell values in Raster[*array*] outside the range 0 to 1 are clipped. ■ If *array* has dimensions {*m*, *n*}, then Raster[*array*] is assumed to occupy the rectangle Rectangle[{0, 0}, {*m*, *n*}]. ■ Raster[*array*, {{*xmin*, *ymin*}, {*xmax*, *ymax*}}] specifies that the raster should be taken instead to fill the rectangle Rectangle[{*xmin*, *ymin*}, {*xmax*, *ymax*}]. ~■ Scaled and Offset can be used to specify the coordinates for the rectangle. ■ Raster[*array*, *rect*, {*zmin*, *zmax*}] specifies that cell values should be scaled so that *zmin* corresponds to 0 and *zmax* corresponds to 1. Cell values outside this range are clipped. ■ See page 477 of the complete *Mathematica Book*. ■ See also: RasterArray, DensityGraphics, GraphicsArray.

+■ Rationals

Rationals represents the domain of rational numbers, as in $x \in$ Rationals.

$x \in$ Rationals evaluates immediately only if x is a numeric quantity. ■ Simplify[*expr* $\in$ Rationals] can be used to try to determine whether an expression corresponds to a rational number. ■ The domain of integers is taken to be a subset of the domain of rationals. ■ Rationals is output in TraditionalForm as $\mathbb{Q}$. ■ See page 54 of this Addendum. ■ See also: Element, Simplify, Algebraics, Integers, Rational, Denominator.

~■ RealDigits

RealDigits[x] gives a list of the digits in the approximate real number x, together with the number of digits that are to the left of the decimal point.

RealDigits[x, b] gives a list of base-b digits in x.

+ RealDigits[x, b, *len*] gives a list of *len* digits.

+ RealDigits[x, b, *len*, n] gives *len* digits starting with the coefficient of b^n.

RealDigits[x] normally returns a list of digits whose length is equal to Precision[x]. +■ RealDigits[x] and RealDigits[x, b] normally require that x be an approximate real number, returned for example by N. RealDigits[x, b, *len*] also works on exact numbers. +■ For exact rational numbers RealDigits[x] returns a list of digits of the form {a_1, a_2, ... , {b_1, b_2, ... }} representing the digits a_i followed by infinite cyclic repetition of the b_i. +■ If *len* is larger than Log[10, b] Precision[x] remaining digits are filled in as Indeterminate. +■ RealDigits[x, b, *len*, n] starts with the digit which is the coefficient of b^n, truncating or padding with zeros as necessary. +■ RealDigits[x, b, *len*, -1] starts with the digit immediately to the right of the base-b decimal point in x. ■ The base b in RealDigits[x, b] need not be an integer. For any real b such that $b > 1$, RealDigits[x, b] successively finds the largest integer multiples of powers of b that can be removed while leaving a non-negative remainder. +■ RealDigits[x] discards the sign of x. +■ FromDigits can be used as the inverse of RealDigits. ■ See page 42 of this Addendum and page 700 of the complete *Mathematica Book*. ■ See also: MantissaExponent, IntegerDigits, BaseForm, FromDigits, ContinuedFraction.

+■ Reals

Reals represents the domain of real numbers, as in $x \in$ Reals.

$x \in$ Reals evaluates immediately only if x is a numeric quantity. ■ Simplify[*expr* $\in$ Reals] can be used to try to determine whether an expression corresponds to a real number. ■ Within Simplify and similar functions, objects that satisfy inequalities are always assumed to be real. ■ Reals is output in TraditionalForm as $\mathbb{R}$. ■ See page 2 of this Addendum. ■ See also: Element, Simplify, Real, Integers, Complexes, Algebraics, ComplexExpand, PowerExpand.

~■ Replace

Replace[*expr*, *rules*] applies a rule or list of rules in an attempt to transform the entire expression *expr*.

₊ Replace[*expr*, *rules*, *levelspec*] applies rules to parts of *expr* specified by *levelspec*.

Examples: Replace[x^2, x^2 -> a] ⟶ a. ■ Replace[x + 1, x -> a] ⟶ 1 + x ■ The rules must be of the form *lhs* -> *rhs* or *lhs* :> *rhs*. ■ A list of rules can be given. The rules are tried in order. The result of the first one that applies is returned. If none of the rules apply, the original *expr* is returned. ■ If the rules are given in nested lists, Replace is effectively mapped onto the inner lists. Thus Replace[*expr*, {{r_{11}, r_{12}}, {r_{21}, ... }, ... }] is equivalent to {Replace[*expr*, {r_{11}, r_{12}}], Replace[*expr*, {r_{21}, ... }], ... }. ■ Delayed rules defined with :> can contain /; conditions. ₊■ Level specifications are described on page 989 of the complete *Mathematica Book*. ₊■ The default value for *levelspec* in Replace is {0}, with Heads -> False. ₊■ Replacements are performed to parts specified by *levelspec* even when those parts have Hold or related wrappers. ■ See page 282 of the complete *Mathematica Book*. ■ See also: Rule, Set, ReplacePart, ReplaceList, StringReplace, PolynomialReduce.

₊■ ShowAutoStyles

ShowAutoStyles is an option for Cell which specifies whether styles that are specified to be automatically used for various syntactic and other constructs should be shown.

The default setting is ShowAutoStyles -> True. ■ Details of automatic styles can be specified in the setting for AutoStyleOptions. ■ For example, unmatched delimiters such as brackets are by default shown in purple. ■ See page 34 of this Addendum. ■ See also: StyleBox, DelimiterFlashTime, ShowCursorTracker.

₊■ ShowCursorTracker

ShowCursorTracker is an option for Cell which specifies whether an elliptical spot should appear momentarily to guide the eye if the cursor position jumps.

The default setting is ShowCursorTracker -> True. ■ Line breaking is normally set up so that small changes to expressions in input cells rarely cause large-scale reformatting; the cursor tracker appears whenever reformatting is required that makes the cursor position jump. ■ The cursor tracker is intended to be sufficiently eye-catching to make the low-level human visual system cause an immediate shift in gaze. ■ See page 34 of this Addendum. ■ See also: DelimiterFlashTime, ShowAutoStyles.

₊■ ShowSelection

ShowSelection is an option for Cell which specifies whether to show the current selection highlighted.

ShowSelection is often set for styles of cells or whole notebooks instead of individual cells. ■ Settings for ShowSelection affect only how the selection is displayed, not where it is or how it works. ■ Setting ShowSelection->False is convenient if you want notebook operations to be performed invisibly. ■ See also: Selectable.

~■ Simplify

Simplify[*expr*] performs a sequence of algebraic transformations on *expr*, and returns the simplest form it finds.

⁺ Simplify[*expr*, *assum*] does simplification using assumptions.

~■ Simplify tries expanding, factoring and doing other transformations on expressions, keeping track of the simplest form obtained. ~■ The following options can be given:

ComplexityFunction	Automatic	how to assess the complexity of each form generated
TimeConstraint	300	for how many seconds to try doing any particular transformation
TransformationFunctions	Automatic	functions to try in transforming the expression
Trig	True	whether to do trigonometric as well as algebraic transformations

⁺■ Assumptions can consist of equations, inequalities, domain assertions such as $x \in$ Integers, and logical combinations of these. ⁺■ Example: Simplify[Sqrt[x^2], x ∈ Reals] ⟶ Abs[x]. ⁺■ Simplify can be used on equations, inequalities and domain assertions. ⁺■ Example: Simplify[x^2 > 3, x > 2] ⟶ True. ⁺■ Objects that are specified as satisfying inequalities are always assumed to be real. ⁺■ Example: Simplify[x ∈ Reals, x > 0] ⟶ True. ⁺■ FullSimplify does more extensive simplification than Simplify. ■ See pages 2 and 50 of this Addendum and pages 67 and 783 of the complete *Mathematica Book*. ■ See also: FullSimplify, Factor, Expand, TrigExpand, PowerExpand, ComplexExpand, Element, FunctionExpand.

~■ Sin

Sin[*z*] gives the sine of *z*.

Mathematical function (see Section A.3.10 of the complete *Mathematica Book*). ■ The argument of Sin is assumed to be in radians. (Multiply by Degree to convert from degrees.) ⁺■ Sin is automatically evaluated when its argument is a simple rational multiple of π; for more complicated rational multiples, FunctionExpand can sometimes be used. ■ See page 731 of the complete *Mathematica Book*. ■ See also: ArcSin, Csc, TrigToExp, TrigExpand.

~■ StringDrop

StringDrop["*string*", *n*] gives "*string*" with its first *n* characters dropped.

StringDrop["*string*", -*n*] gives "*string*" with its last *n* characters dropped.

StringDrop["*string*", {*n*}] gives "*string*" with its nth character dropped.

StringDrop["*string*", {*m*, *n*}] gives "*string*" with characters *m* through *n* dropped.

StringDrop uses the standard *sequence specification* (see page 80). ■ Example: StringDrop["abcdefgh", 2] ⟶ cdefgh. ⁺■ StringDrop["*string*", {*m*, *n*, *s*}] drops characters *m* through *n* in steps of *s*. ■ See page 386 of the complete *Mathematica Book*. ■ See also: Drop, StringTake, StringPosition, StringReplacePart.

~■ StringReplace

StringReplace["*string*", "s_1" -> "sp_1"] or
StringReplace["*string*", {"s_1" -> "sp_1", "s_2" -> "sp_2", ... }] replaces the "s_i" by "sp_i" whenever they appear as substrings of "*string*".

StringReplace goes through a string, testing substrings that start at each successive character position. On each substring, it tries in turn each of the transformation rules you have specified. If any of the rules apply, it replaces the substring, then continues to go through the string, starting at the character position after the end of the substring. ⁺■ Delayed replacements of the form "*s*" :> *expr* can be given, so long as *expr* evaluates to a string every time the replacement is used. ■ Setting the option IgnoreCase -> True makes StringReplace treat lower- and upper-case letters as equivalent. ■ See page 389 of the complete *Mathematica Book*. ■ See also: Replace, StringReplacePart, StringMatchQ, StringPosition, ToLowerCase, ToUpperCase.

■ StringTake

StringTake["*string*", *n*] gives a string containing the first *n* characters in "*string*".

StringTake["*string*", -*n*] gives the last *n* characters in "*string*".

StringTake["*string*", {*n*}] gives the *n*th character in "*string*".

StringTake["*string*", {*m*, *n*}] gives characters *m* through *n* in "*string*".

StringTake uses the standard *sequence specification* (see page 80). ■ Example: StringTake["abcdefg", 3] ⟶ abc.
₊■ StringTake["*string*", {*m*, *n*, *s*}] gives characters *m* through *n* in steps of *s*. ■ See page 386 of the complete *Mathematica Book*. ■ See also: Take, StringDrop, StringPosition.

₊■ StruveH

StruveH[*n*, *z*] gives the Struve function $\mathbf{H}_n(z)$.

Mathematical function (see Section A.3.10 of the complete *Mathematica Book*). ■ $\mathbf{H}_n(z)$ for integer *n* satisfies the differential equation $z^2 y'' + zy' + (z^2 - n^2)y = \frac{2}{\pi} \frac{z^{n+1}}{(2n-1)!!}$. ■ StruveH[*n*, *z*] has a branch cut discontinuity in the complex *z* plane running from $-\infty$ to 0. ■ See page 47 of this Addendum. ■ See also: StruveL, BesselJ.

₊■ StruveL

StruveL[*n*, *z*] gives the modified Struve function $\mathbf{L}_n(z)$.

Mathematical function (see Section A.3.10 of the complete *Mathematica Book*). ■ $\mathbf{L}_n(z)$ for integer *n* is related to the ordinary Struve function by $\mathbf{L}_n(iz) = -ie^{-in\pi/2}\mathbf{H}_n(z)$. ■ StruveH[*n*, *z*] has a branch cut discontinuity in the complex *z* plane running from $-\infty$ to 0. ■ See page 47 of this Addendum. ■ See also: StruveH, BesselJ.

₊■ Subresultants

Subresultants[*poly*₁, *poly*₂, *var*] generates a list of the principal subresultant coefficients of the polynomials *poly*₁ and *poly*₂ with respect to the variable *var*.

The first *k* subresultants of two polynomials *a* and *b*, both with leading coefficient one, are zero when *a* and *b* have *k* common roots. ■ Subresultants returns a list whose length is
Min[Exponent[*poly*₁, *var*], Exponent[*poly*₂, *var*]] + 1. ■ See page 49 of this Addendum. ■ See also: Resultant, PolynomialGCD, Eliminate, Minors.

◼ Take

Take[*list*, *n*] gives the first *n* elements of *list*.

Take[*list*, -*n*] gives the last *n* elements of *list*.

Take[*list*, {*m*, *n*}] gives elements *m* through *n* of *list*.

✦ Take[*list*, {*m*, *n*, *s*}] gives elements *m* through *n* in steps of *s*.

✦ Take[*list*, *seq*$_1$, *seq*$_2$, ...] gives a nested list in which elements specified by *seq*$_i$ are taken at level *i* in *list*.

Take uses the standard *sequence specification* (see page 80). ▪ Examples: Take[{a,b,c,d,e}, 3] ⟶ {a, b, c}. ▪ Take[{a,b,c,d,e}, -2] ⟶ {d, e}. ✦▪ Take[Range[15], {3, 12, 4}] ⟶ {3, 7, 11}. ▪ Take can be used on an object with any head, not necessarily List. ✦▪ Take[*list*, *seq*$_1$, *seq*$_2$] effectively extracts a submatrix from *list*. ✦▪ Example: Take[{{a,b,c},{d,e,f}}, -1, 2] ⟶ {{d, e}}. ▪ See page 7 of this Addendum and page 118 of the complete *Mathematica Book*. ▪ See also: Part, Drop, StringTake, Select, Cases, Partition, PadLeft. ▪ Related package: LinearAlgebra`MatrixManipulation`.

✦◼ Tr

Tr[*list*] finds the trace of the matrix or tensor *list*.

Tr[*list*, *f*] finds a generalized trace, combining terms with *f* instead of Plus.

Tr[*list*, *f*, *n*] goes down to level *n* in *list*.

Tr[*list*] sums the diagonal elements *list*[[*i*, *i*, ...]]. ▪ Tr works for rectangular as well as square matrices and tensors. ▪ See page 64 of this Addendum. ▪ See also: Transpose, Det, DiagonalMatrix, Eigenvalues.

✦◼ TransformationFunctions

TransformationFunctions is an option for Simplify and FullSimplify which gives the list of functions to apply to try to transform parts of an expression.

The default setting TransformationFunctions->Automatic uses a built-in collection of transformation functions. ▪ TransformationFunctions->{*f*$_1$, *f*$_2$, ... } uses only the functions *f*$_i$. ▪ TransformationFunctions->{Automatic, *f*$_1$, *f*$_2$, ... } uses built-in transformation functions together with the functions *f*$_i$. ▪ See also: Simplify, FullSimplify, ReplaceAll, ExcludedForms, FunctionExpand.

✦◼ UnitStep

UnitStep[*x*] represents the unit step function, equal to 0 for $x < 0$ and 1 for $x > 0$.

UnitStep[x_1, x_2, ...] represents the multidimensional unit step function which is 0 unless all the x_i are positive.

Some transformations are done automatically when UnitStep appears in a product of terms. ▪ UnitStep provides a convenient way to represent piecewise continuous functions. ▪ UnitStep has attribute Orderless. ▪ For exact numeric quantities, UnitStep internally uses numerical approximations to establish its result. This process can be affected by the setting of the global variable $MaxExtraPrecision. ▪ See page 59 of this Addendum. ▪ See also: Sign, Positive, DiracDelta, DiscreteDelta, KroneckerDelta.

■ Zeta

Zeta[*s*] gives the Riemann zeta function $\zeta(s)$.

Zeta[*s*, *a*] gives the generalized Riemann zeta function $\zeta(s, a)$.

Mathematical function (see Section A.3.10 of the complete *Mathematica Book*). ■ $\zeta(s) = \sum_{k=1}^{\infty} k^{-s}$. ■ $\zeta(s, a) = \sum_{k=0}^{\infty} (k + a)^{-s}$, where any term with $k + a = 0$ is excluded. ■ Zeta[*s*] has no branch cut discontinuities. +■ FullSimplify and FunctionExpand include transformation rules for Zeta. ■ See page 46 of this Addendum and page 742 of the complete *Mathematica Book*. ■ Implementation notes: see page 83. ■ See also: PolyLog, HarmonicNumber, LerchPhi, RiemannSiegelZ, StieltjesGamma, Glaisher, PrimePi. ■ Related package: NumberTheory`Ramanujan`.

■ ZTransform

ZTransform[*expr*, *n*, *z*] gives the Z transform of *expr*.

The Z transform of a function $f(n)$ is defined to be $\sum_{n=0}^{\infty} f(n) z^{-n}$. ■ See page 58 of this Addendum. ■ See also: InverseZTransform, LaplaceTransform, Sum, Series. ■ Related package: DiscreteMath`RSolve`.

■ $ByteOrdering

$ByteOrdering gives the native ordering of bytes in binary data on your computer system.

Possible values of $ByteOrdering are +1 and -1. ■ +1 corresponds to big endian (appropriate for 680x0 and many other processors); -1 corresponds to little endian (appropriate for x86 processors). ■ +1 corresponds to having the most significant byte first; -1 to having the least significant byte first. ■ +1 is the order obtained from IntegerDigits[*n*, 256]. ■ $ByteOrdering gives the default setting for the ByteOrdering option in Import and Export. ■ See also: $ProcessorType.

■ $ExportFormats

$ExportFormats gives a list of export formats currently supported in your *Mathematica* system.

The strings that appear in $ExportFormats are the possible third arguments to Export. ■ See also: $ImportFormats, Export, $Packages.

■ $ImportFormats

$ImportFormats gives a list of import formats currently supported in your *Mathematica* system.

The strings that appear in $ImportFormats are the possible second arguments to Import. ■ See also: $ExportFormats, Import, $Packages.

■ $MaxPrecision

$MaxPrecision gives the maximum number of digits of precision to be allowed in arbitrary-precision numbers.

The default value of $MaxPrecision is 1000000. ■ $MaxPrecision = Infinity uses the maximum value possible on a particular computer system, given roughly by Log[10, $MaxNumber]. ■ $MaxPrecision is measured in decimal digits, and need not be an integer. ■ See page 709 of the complete *Mathematica Book*. ■ See also: $MinPrecision, $MaxExtraPrecision.

A.13 Incompatible Changes since Version 2

A very large number of new capabilities have been added since Version 2 of *Mathematica*. Functions that were added in Version 3.0 are indicated by +■ in the listing of built-in objects that begins on page 89; those that were added in Version 4 are indicated by +■.

Sections in the main part of this book that were completely new for Version 3.0 are indicated by +■; those new for Version 4 are indicated by +■. Those that were substantially revised or enhanced for Version 3.0 are indicated by ~■; those substantially revised or enhanced for Version 4 are indicated by ~■. Note that even though no significant change may have been made in the description of particular capabilities, the capabilities themselves have in many cases been greatly extended.

For the most part, the changes made in Versions 3.0 and 4 are purely additions. There are, however, a few features of Versions 3.0 and 4 which are incompatible with earlier versions.

- •
- •
- •

For Version 4 these include:

- N[0] now yields a machine-precision zero rather than an exact zero.

- FullOptions has been superseded by AbsoluteOptions, which yields results in the same form as Options.

- Element[x, y] or $x \in y$ now has built-in evaluation rules.

- A new second argument has been added to CompiledFunction to allow easier manipulation and composition of compiled functions.

In general, great effort is made to ensure that successive versions of *Mathematica* are fully compatible. However, in maintaining the coherence of the system, certain functions must gradually be changed. In Version 3.0, functions that became obsolete in Version 2.0 are no longer supported at all.

*A.14 Developer Context Objects in Mathematica 4

The objects listed below are included in the `Developer`` context in *Mathematica* Version 4; some of them may change or be moved to the `System`` context in subsequent versions of *Mathematica*.

The `Developer`` context includes functions that directly access specific internal algorithms and capabilities of *Mathematica* that are normally used only as part of more general functions.

`Developer``*name*	access a specific object in the `Developer`` context
`<<Developer``	set up to be able to access all `Developer`` objects by name

Ways to access functions in the `Developer`` context.

Note that `<<Developer`` adds the `Developer`` context to your `$ContextPath`. You can remove it again by explicitly modifying the value of `$ContextPath`. You can set up to access `Developer`` context objects automatically in all *Mathematica* sessions by adding `<<Developer`` to your `init.m` file.

■ Developer`BesselSimplify

`BesselSimplify[`*expr*`]` transforms Bessel functions in *expr*, trying to either decrease the number of Bessel functions, or convert Bessel functions into more elementary functions.

`BesselSimplify` is automatically used inside `FullSimplify` and `FunctionExpand`.

■ Developer`BitLength

`BitLength[`n`]` gives the number of binary bits necessary to represent the integer n.

For positive n, `BitLength[`n`]` is effectively an efficient version of `Ceiling[Log[2, `n`]]`. ■ For negative n it is equivalent to `BitLength[BitNot[`n`]]`. ■ n must be non-negative. ■ See also: `IntegerExponent`, `MantissaExponent`.

■ Developer`BitShiftLeft

`BitShiftLeft[`n`, `d`]` shifts the binary bits in the integer n to the left by d places, padding with zeros on the right.

`BitShiftLeft[`n`, `d`]` is equivalent to $n\ 2^d$. ■ Negative values of d shift to the right.

■ Developer`BitShiftRight

`BitShiftRight[`n`, `d`]` shifts the binary bits in the integer n to the right by d places, dropping bits that are shifted past the units position on the right.

`BitShiftRight[`n`, `d`]` is equivalent to `IntegerPart[`$n/2^d$`]`. ■ Negative values of d shift to the left. ■ See also: `MantissaExponent`.

■ **Developer`CharacteristicPolynomial**

CharacteristicPolynomial[*m*, *x*] computes the characteristic polynomial in the variable *x* for a matrix *m*.

The matrix *m* can contain approximate or exact numbers, or symbolic expressions. ■ See also: Det.

■ **Developer`ClearCache**

ClearCache[ ] clears internal caches of stored results.

ClearCache is useful if one needs to generate worst-case timing results independent of previous computations. ■ ClearCache["Numeric"] clears only caches of numeric results. ■ ClearCache["Symbolic"] clears only caches of symbolic results. ■ See also: Update.

■ **Developer`ClipboardNotebook**

ClipboardNotebook[ ] gives the notebook object corresponding to the invisible notebook corresponding to the clipboard for copy and paste operations.

It is possible to both read and write to the clipboard notebook. ■ See also: NotebookWrite, SelectedNotebook, EvaluationNotebook.

■ **Developer`FibonacciSimplify**

FibonacciSimplify[*expr*, *assum*] tries to simplify combinations of symbolic Fibonacci numbers in *expr* using assumptions *assum*.

Example:
Developer`FibonacciSimplify[Fibonacci[n-1]+Fibonacci[n-2], Element[n, Integers]] ⟶ Fibonacci[n].
■ FibonacciSimplify can typically perform transformations only when arguments of Fibonacci numbers are specified to be integers. ■ FibonacciSimplify is automatically used inside FullSimplify and FunctionExpand.

■ **Developer`FileInformation**

FileInformation["*name*"] gives information about the file with the specified name.

FileInformation gives a list of rules which include information on the name, size and modification time of a file. ■ ToDate can be used to convert modification times to dates. ■ See also: FileNames, FileDate, FileByteCount, AbsoluteOptions, Developer`NotebookInformation.

■ **Developer`FromPackedArray**

FromPackedArray[*expr*] unpacks *expr* so that its internal representation is not a packed array.

Using FromPackedArray will not change results generated by *Mathematica*, but can reduce speed of execution and increase memory usage. ■ If *expr* is not a packed array, FromPackedArray[*expr*] returns *expr* unchanged. ■ See also: Developer`ToPackedArray, Developer`PackedArrayQ, ByteCount.

■ **Developer`GammaSimplify**

GammaSimplify[*expr*] transforms gamma functions in *expr*, trying to either decrease the number of gamma functions, or convert combinations of them into more elementary functions.

GammaSimplify is automatically used inside FullSimplify and FunctionExpand.

■ **Developer`HelpBrowserNotebook**

HelpBrowserNotebook[ ] gives the notebook object corresponding to the notebook portion of the Help Browser window.

See also: Developer`MessagesNotebook, SelectedNotebook, EvaluationNotebook.

■ **Developer`HermiteNormalForm**

HermiteNormalForm[m] gives the Hermite normal form of an integer matrix m.

The result is given in the form $\{u, r\}$ where u is a unimodular matrix, r is an upper triangular matrix, and u . m == r. ■ See also: RowReduce, LatticeReduce.

■ **Developer`HessenbergDecomposition**

HessenbergDecomposition[m] gives the Hessenberg decomposition of a matrix m.

The result is given in the form $\{p, h\}$ where p is a unitary matrix such that p . h . Conjugate[Transpose[p]] == m. ■ See also: SchurDecomposition.

■ **Developer`InequalityInstance**

InequalityInstance[*ineqs*, $\{x_1, x_2, \dots \}$] gives, if possible, a set of real values of the x_i for which the inequalities *ineqs* are satisfied.

InequalityInstance gives results as a list of rules, in the same format as used by Solve. ■ If there are no choices of the x_i for which the inequalities are satisfied, InequalityInstance returns { }. ■ InequalityInstance effectively returns an arbitrary point in the region defined by *ineqs*. ■ For given *ineqs*, the point returned will always be the same. ■ The *ineqs* can contain logical combinations or a list of inequalities, equations and inequations.

■ **Developer`MachineIntegerQ**

MachineIntegerQ[*expr*] returns True if *expr* corresponds to a machine-sized integer, and False otherwise.

On a typical computer system machine-sized integers must lie in the range $-2^{31} + 1$ to $+2^{31} - 1$ or $-2^{63} + 1$ to $+2^{63} - 1$. ■ Results from *Mathematica* are not affected by whether an integer is machine-sized or not; the speed of operations may however be affected. ■ See also: Developer`$MaxMachineInteger, Precision.

■ **Developer`MessagesNotebook**

MessagesNotebook[ ] gives the notebook to which messages generated by the notebook front end will be sent.

MessagesNotebook returns a NotebookObject. ■ See also: SelectedNotebook, EvaluationNotebook.

■ **Developer`NotebookConvert**

NotebookConvert["*name*"] converts a *Mathematica* notebook from a previous version of *Mathematica* to one for the current version.

NotebookConvert takes a file named *name*.ma or *name* and generates a file named *name*.nb. ■ Notebook conversion is done automatically by the front end if you try to open a Version 2 notebook. ■ See also: NotebookGet.

■ Developer`NotebookInformation

NotebookInformation[*obj*] gives information about the notebook represented by the specified notebook object.

NotebookInformation gives a list of rules which include information on times when data associated with the notebook was most recently modified. ■ ToDate can be used to convert these times to dates. ■ See also: AbsoluteOptions, Developer`FileInformation.

■ Developer`PackedArrayForm

PackedArrayForm[*expr*] prints with packed arrays in *expr* shown in summary form, without all of their elements explicitly given.

With a notebook front end, packed arrays are typically output as cyan-colored boxes. ■ See also: Short, Shallow.

■ Developer`PackedArrayQ

PackedArrayQ[*expr*] returns True if *expr* is a packed array in its internal representation, and returns False otherwise.

PackedArrayQ[*expr*, *type*] returns True if *expr* is a packed array of objects of the specified type.

PackedArrayQ[*expr*, *type*, *rank*] returns True if *expr* is a packed array of the specified rank.

Supported types are Integer, Real and Complex. ■ See also: Developer`ToPackedArray, ByteCount.

■ Developer`PolyGammaSimplify

PolyGammaSimplify[*expr*] transforms polygamma functions in *expr*, trying to either decrease the number of polygamma functions, or convert combinations of them into more elementary functions.

PolyGammaSimplify is automatically used inside FullSimplify and FunctionExpand.

■ Developer`PolyLogSimplify

PolyLogSimplify[*expr*] transforms polylogarithm functions in *expr*, trying to either decrease the number of polylogarithm functions, or convert combinations of them into more elementary functions.

PolyLogSimplify is automatically used inside FullSimplify and FunctionExpand.

■ Developer`PolynomialDivision

PolynomialDivision[*p*, *q*, *x*] gives a list of the quotient and remainder of *p* and *q*, treated as polynomials in *x*.

The remainder will always have a degree not greater than *q*. ■ See also: PolynomialQuotient, PolynomialRemainder, PolynomialReduce.

■ **`Developer`ReplaceAllUnheld`**

`ReplaceAllUnheld[`*expr*, *rules*`]` applies a rule or list of rules in an attempt to transform each subpart of *expr* that would be automatically evaluated.

`ReplaceAll` operates on all subparts of an expression; `ReplaceAllUnheld` operates only on those subparts that would normally be evaluated. ■ Example: `Developer`ReplaceAllUnheld[If[a, a, a], a->b]` ⟶ `If[b, a, a]`. ■ See also: `ReplaceAll`, `Hold`, `Verbatim`.

■ **`Developer`SetSystemOptions`**

`SetSystemOptions["`*name*`"->`*value*`]` sets a specified system option.

See also: `Developer`SystemOptions`.

■ **`Developer`SparseLinearSolve`**

`SparseLinearSolve[`*smat*, *vec*`]` solves a sparse linear system; the matrix *smat* is represented in the form $\{\{i_1, j_1\}$->$a_1, \{i_2, j_2\}$->$a_2, \ldots \}$, so that the element at position i_k, j_k has value a_k and all unspecified elements are taken to be zero.

The vector *vec* should be a list of numbers. ■ The result from `SparseLinearSolve` is a list of numbers of the same length as *vec*. ■ The matrix *smat* is effectively assumed to be square. ■ The elements of *smat* do not need to have head `Rule`; they can also for example have head `List`. ■ The functionality of `SparseLinearSolve` is automatically accessed when you apply `Solve` or `NSolve` to a sparse system of linear equations given in symbolic form. ■ See also: `LinearSolve`.

■ **`Developer`SystemOptions`**

`SystemOptions[ ]` gives the current settings for all system options.

`SystemOptions["`*name*`"]` gives the current setting for the system option with the specified name.

System options specify internal parameters relevant to the operation of *Mathematica* on particular computer systems. ■ See also: `Developer`SetSystemOptions`.

■ **`Developer`ToPackedArray`**

`ToPackedArray[`*expr*`]` uses packed arrays if possible in the internal representation of *expr*.

Using `ToPackedArray` will not change results generated by *Mathematica*, but can enhance speed of execution and reduce memory usage. ■ `ToPackedArray` is effectively used automatically by many functions that generate large lists. ■ `ToPackedArray` will successfully pack full lists of any depth containing machine-sized integers and machine-sized approximate real and complex numbers. ■ `ToPackedArray[`*expr*, *type*`]` will when possible convert entries in *expr* to be of the specified type. ■ Possible types are: `Integer`, `Real` and `Complex`. ■ Only machine-sized numbers can be stored in packed form. ■ The option `Tolerance->`*tol* can be used to specify when small numerical values can be ignored in conversion to more restrictive types, and when they must prevent conversion to packed form. ■ See also: `Developer`FromPackedArray`, `ByteCount`.

■ **`Developer`TrigToRadicals`**

`TrigToRadicals[`*expr*`]` converts trigonometric functions to radicals whenever possible in *expr*.

`TrigToRadicals` operates on trigonometric functions whose arguments are rational multiples of π. ■ `TrigToRadicals` is automatically used inside `FullSimplify` and `FunctionExpand`.

■ **Developer'ZeroQ**

ZeroQ[*expr*] returns True if built-in transformations allow it to be determined that *expr* is numerically equal to zero, and returns False otherwise.

ZeroQ uses a combination of symbolic transformations and randomized numerical evaluation. ■ If ZeroQ[*expr*] returns False it does not necessarily mean that *expr* is mathematically not equal to zero; all it means is that built-in transformations did not allow this to be determined. ■ See also: Experimental'ImpliesQ.

■ **Developer'ZetaSimplify**

ZetaSimplify[*expr*] transforms zeta functions in *expr*, trying to either decrease the number of zeta functions, or convert combinations of them into more elementary functions.

ZetaSimplify is automatically used inside FullSimplify and FunctionExpand.

■ **Developer'$MaxMachineInteger**

$MaxMachineInteger gives the maximum integer that is represented internally as a single atomic data element on your computer system.

$MaxMachineInteger is typically $2^{n-1} - 1$ on a computer system that is referred to as an *n*-bit system. ■ Arithmetic operations involving integers smaller than $MaxMachineInteger are typically faster than those involving larger integers. ■ See also: Developer'MachineIntegerQ, $MaxMachineNumber, $MachinePrecision.

+A.15 Experimental Context Objects in *Mathematica* 4

The objects listed below are experimental in *Mathematica* Version 4, and are subject to change in subsequent versions of *Mathematica*.

`Experimental`*name*	access a specific object in the `Experimental`` context
`<<Experimental``	set up to be able to access all `Experimental`` objects by name

Ways to access functions in the `Experimental`` context.

Note that `<<Experimental`` adds the `Experimental`` context to your `$ContextPath`. You can remove it again by explicitly modifying the value of `$ContextPath`. You can set up to access `Experimental`` context objects automatically in all *Mathematica* sessions by adding `<<Experimental`` to your `init.m` file.

`<<RealTime3D``	set up real-time 3D graphics in the notebook front end
`<<Default3D``	revert to using default 3D graphics rendering

An additional experimental feature under Microsoft Windows.

■ Experimental`BinaryExport

BinaryExport[*channel*, *expr*, *format*] exports *expr* to *channel* as binary data in the specified format.

The basic elements that can appear in the format specification are:

"Byte"	8-bit unsigned integer
"Character8"	8-bit character
"Character16"	16-bit character
"Complex64"	IEEE single-precision complex number
"Complex128"	IEEE double-precision complex number
"Integer8"	8-bit signed integer
"Integer16"	16-bit signed integer
"Integer32"	32-bit signed integer
"Integer64"	64-bit signed integer
"Real32"	IEEE single-precision real number
"Real64"	IEEE double-precision real number
"Real128"	IEEE quadruple-precision real number
"TerminatedString"	null-terminated string of 8-bit characters
"UnsignedInteger8"	8-bit unsigned integer
"UnsignedInteger16"	16-bit unsigned integer
"UnsignedInteger32"	32-bit unsigned integer
"UnsignedInteger64"	64-bit unsigned integer

■ These elements can be combined in lists or other expressions. ■ The pattern *format*.. represents a sequence of one or more copies of a format. ■ Example: {"Byte"..} represents a list of one or more bytes.
■ {"Integer32", "Real32"}.. represents a list of one or more repetitions of a 32-bit integer followed by a single-precision real. ■ The expressions *format* and *expr* are assumed to have the same structure, except for the replacement of *patt*.. by explicit sequences. ■ Elements are sent as exported data in the order that they would be accessed by a function such as MapAll. ■ BinaryExport coerces data to correspond to the format specifications given. ■ Integers that do not fit have their high-order bits dropped. ■ The channel used in BinaryExport can be a file specified by its name, a pipe or an OutputStream. ■ Under Microsoft Windows, the output stream must have been opened with DOSTextFormat->False. ■ When BinaryExport exports data to an output stream, it leaves the stream position directly after what it has exported. ■ If BinaryExport opens a file or pipe, it closes it again when it is finished. ■ The following options can be given:

ByteOrdering	Automatic	what byte ordering to assume
CharacterEncoding	Automatic	what encoding to use for characters

■ See also: Experimental`BinaryImport, Experimental`BinaryExportString, Export, FromCharacterCode.
■ Note: this is an experimental feature, and in future versions of *Mathematica* it may not be supported, or may have a different specification.

■ Experimental`BinaryExportString

BinaryExportString[*expr*, *format*] returns a string corresponding to *expr* exported as binary data.

See notes for BinaryExport.

■ Experimental`BinaryImport

BinaryImport[*channel*, *format*] imports binary data from *channel* in the specified format.

The basic elements that can appear in the format specification are:

"Byte"	8-bit unsigned integer
"Character8"	8-bit character
"Character16"	16-bit character
"Complex64"	IEEE single-precision complex number
"Complex128"	IEEE double-precision complex number
"Integer8"	8-bit signed integer
"Integer16"	16-bit signed integer
"Integer32"	32-bit signed integer
"Integer64"	64-bit signed integer
"Real32"	IEEE single-precision real number
"Real64"	IEEE double-precision real number
"Real128"	IEEE quadruple-precision real number
"TerminatedString"	null-terminated string of 8-bit characters
"UnsignedInteger8"	8-bit unsigned integer
"UnsignedInteger16"	16-bit unsigned integer
"UnsignedInteger32"	32-bit unsigned integer
"UnsignedInteger64"	64-bit unsigned integer

■ These elements can be combined in lists or other expressions. ■ The pattern *format*.. represents a sequence of one or more copies of a format. ■ Example: {"Byte"..} represents a list of one or more bytes.
■ {"Integer32", "Real32"}.. represents a list of one or more repetitions of a 32-bit integer followed by a single-precision real. ■ BinaryImport returns an object in which each element of the format specification has been replaced by imported data. ■ Numerical elements are returned as *Mathematica* numbers; character and string elements are returned as *Mathematica* strings. ■ Elements in a format specification are filled from imported data in the order that they would be accessed by a function such as MapAll. ■ The channel used in BinaryImport can be a file specified by its name, a pipe or an InputStream. ■ Under Microsoft Windows, the input stream must have been opened with DOSTextFormat->False. ■ When BinaryImport imports data from an input stream, it leaves the stream position directly after what it has imported. ■ If BinaryImport opens a file or pipe, it closes it again when it is finished. ■ The following options can be given:

ByteOrdering	Automatic	what byte ordering to assume
CharacterEncoding	Automatic	what encoding to use for characters
Path	$Path	the path to search for files

■ See also: Experimental`BinaryExport, Experimental`BinaryImportString, Import, ToCharacterCode. ■ Note: this is an experimental feature, and in future versions of *Mathematica* it may not be supported, or may have a different specification.

■ Experimental`BinaryImportString

BinaryImportString["*string*", *format*] imports binary data from a string in the specified format.

See notes for BinaryImport.

■ Experimental`CholeskyDecomposition

CholeskyDecomposition[*m*] computes the Cholesky decomposition of a matrix *m*.

CholeskyDecomposition[*m*] returns a list of the form {*lmat*, *perm*, *diag*}, where *lmat* is a lower-triangular matrix, *perm* is a permutation vector and *diag* is a vector corresponding to the leading diagonal of a matrix. ■ When *perm* is the identity permutation and *diag* is a zero vector, then *lmat* . Transpose[*lmat*] is exactly the original matrix *m*. ■ In general, *lmat* . Transpose[*lmat*] is given by Transpose[*p*] . *m* . *p* + DiagonalMatrix[*diag*] where *p* = IdentityMatrix[Length[*perm*]][[*perm*]]. ■ CholeskyDecomposition works with both numerical and symbolic square matrices. ■ CholeskyDecomposition regularizes all Hermitian numerical matrices to make them positive definite Hermitian. ■ See also: LUDecomposition, LUBackSubstitution, LinearSolve. ■ Note: this is an experimental feature, and in future versions of *Mathematica* it may not be supported, or may have a different specification.

■ Experimental`CompileEvaluate

CompileEvaluate[*expr*] compiles *expr* and then evaluates the resulting compiled code.

CompileEvaluate[*expr*] always evaluates to the same result as *expr* alone, but is faster for certain types of expressions, particularly ones representing large numerical computations. ■ See also: Compile. ■ Note: this is an experimental feature, and in future versions of *Mathematica* it may not be supported, or may have a different specification.

■ Experimental`CylindricalAlgebraicDecomposition

CylindricalAlgebraicDecomposition[*ineqs*, {x_1, x_2, ... }] finds a decomposition of the region represented by the inequalities *ineqs* into cylindrical parts whose directions correspond to the successive x_i.

Example: Experimental`CylindricalAlgebraicDecomposition[x^2 + y^2 < 1, {x, y}] ⟶ $-1 < x < 1$ && $-\sqrt{1-x^2} < y < \sqrt{1-x^2}$. ■ CylindricalAlgebraicDecomposition assumes that all variables are real. ■ Lists or logical combinations of inequalities can be given. ■ CylindricalAlgebraicDecomposition returns inequalities whose bounds in general involve algebraic functions. ■ The algorithms of CylindricalAlgebraicDecomposition are used automatically inside FullSimplify and Resolve. ■ See also: Experimental`ExistsRealQ, Experimental`ForAllRealQ, Experimental`ImpliesRealQ, Experimental`Resolve, Experimental`Minimize. ■ Note: this is an experimental feature, and in future versions of *Mathematica* it may not be supported, or may have a different specification.

■ Experimental`ExistsRealQ

ExistsRealQ[*ineqs*, {x_1, x_2, ... }] tests whether there exist real values of the x_i for which the inequalities and equations *ineqs* are satisfied.

See also: Experimental`ForAllRealQ, Experimental`ImpliesRealQ, Developer`InequalityInstance, Eliminate, Experimental`CylindricalAlgebraicDecomposition. ■ Note: this is an experimental feature, and in future versions of *Mathematica* it may not be supported, or may have a different specification.

■ Experimental`ExtendedLinearSolve

ExtendedLinearSolve[*m*] is an extended version of LinearSolve[*m*, *v*] which supports the one-argument form ExtendedLinearSolve[*m*]. This one-argument form yields a LinearSolveFunction object which when applied to a vector *v* gives the same result as LinearSolve[*m*, *v*].

ExtendedLinearSolve is particularly efficient when repeated solutions are needed for several different vectors *v*. ■ The matrix *m* can be given in the sparse form used by Developer`SparseLinearSolve. ■ See also: LinearSolve, Inverse, LUDecomposition, Developer`SparseLinearSolve. ■ Note: this is an experimental feature, and in future versions of *Mathematica* it may not be supported, or may have a different specification.

■ Experimental`ExtendedLUBackSubstitution

ExtendedLUBackSubstitution[*data*, *b*, *f*] is an extended version of LUBackSubstitution[*data*, *b*] which allows the function *f* to be effectively applied to the original matrix.

The possible values for *f* are Identity, Transpose and Conjugate.

The matrix *m* can be given in the sparse form used by Developer`SparseLinearSolve. ■ See also: LUBackSubstitution, Developer`SparseLinearSolve. ■ Note: this is an experimental feature, and in future versions of *Mathematica* it may not be supported, or may have a different specification.

■ Experimental`FileBrowse

FileBrowse[ ] brings up a file browser to pick the name of a file.

FileBrowse returns as a string the absolute name of the file picked. ■ FileBrowse["*name*"] brings up a file browser with the specified default file name. ■ FileBrowse["*name*", "*directory*"] brings up a file browser starting in the specified directory. ■ FileBrowse can be used to find names of files for both reading and writing. ■ FileBrowse is intended primarily for use with local kernels. ■ See also: FileNames, Get, Put. ■ Note: this is an experimental feature, and in future versions of *Mathematica* it may not be supported, or may have a different specification.

■ Experimental`ForAllRealQ

ForAllRealQ[*ineqs*, {x_1, x_2, ... }] tests whether for all real values of the x_i the inequalities and equations *ineqs* are satisfied.

See also: Experimental`ExistsRealQ, Experimental`ImpliesRealQ, SolveAlways, Experimental`CylindricalAlgebraicDecomposition. ■ Note: this is an experimental feature, and in future versions of *Mathematica* it may not be supported, or may have a different specification.

■ Experimental`ImpliesQ

ImpliesQ[$expr_1$, $expr_2$] tests whether the expression $expr_1$ implies $expr_2$.

ImpliesQ returns False if it cannot determine whether $expr_1$ implies $expr_2$, using any of its built-in transformation rules. ■ The related function Implies[$expr_1$, $expr_2$] remains unevaluated if it cannot immediately determine whether $expr_1$ implies $expr_2$. ■ See also: Implies, Experimental`ImpliesRealQ, FullSimplify. ■ Note: this is an experimental feature, and in future versions of *Mathematica* it may not be supported, or may have a different specification.

■ Experimental`ImpliesRealQ

ImpliesRealQ[*ineqs*₁, *ineqs*₂] tests whether the inequalities and equations *ineqs*₁ imply the *ineqs*₂ for all real values of all variables.

See also: Experimental`ForAllRealQ, Experimental`ExistsRealQ, Experimental`ImpliesQ, Experimental`CylindricalAlgebraicDecomposition. ■ Note: this is an experimental feature, and in future versions of *Mathematica* it may not be supported, or may have a different specification.

■ Experimental`Minimize

Minimize[*expr*, *cons*, {x_1, x_2, ... }] minimizes *expr* with respect to the real variables x_i subject to the constraints *cons*.

The constraints can involve both equations and inequalities. ■ Minimize returns both the minimum value and the position of the minimum, in the same form as ConstrainedMin and FindMinimum. ■ See also: ConstrainedMin, FindMinimum, Experimental`CylindricalAlgebraicDecomposition. ■ Note: this is an experimental feature, and in future versions of *Mathematica* it may not be supported, or may have a different specification.

■ Experimental`Resolve

Resolve[*expr*] attempts to eliminate quantifiers in *expr*.

Resolve handles quantifiers of the form Exists[*x*, *condition*, *expr*] and ForAll[*x*, *condition*, *expr*]. ■ Exists[*x*, *c*, *expr*] can be input in the form $\exists_{x,c}$ *expr*. ■ ForAll[*x*, *c*, *expr*] can be input in the form $\forall_{x,c}$ *expr*. ■ Lists of variables can be used in Exists and ForAll. ■ The algorithms of Resolve are used automatically inside FullSimplify. ■ See also: Experimental`CylindricalAlgebraicDecomposition, FullSimplify. ■ Note: this is an experimental feature, and in future versions of *Mathematica* it may not be supported, or may have a different specification.

■ Experimental`ValueFunction

ValueFunction[*symb*] represents a function to be applied whenever the symbol *symb* gets a new value.

The assignment ValueFunction[*symb*] = *f* specifies that whenever *symb* gets a new value *val*, the expression *f*[*symb*, *val*] should be evaluated. ■ If the value of *symb* is cleared, *f*[*symb*] is evaluated. ■ ValueFunction takes account of all ways that the value of a symbol can be changed, not just Set. ■ See also: Trace. ■ Note: this is an experimental feature, and in future versions of *Mathematica* it may not be supported, or may have a different specification.

■ Experimental`$EqualTolerance

$EqualTolerance gives the number of decimal digits by which two numbers can disagree and still be considered equal according to Equal.

The default setting is equal to Log[10, 2^7], corresponding to a tolerance of 7 binary digits. ■ See also: $MachineEpsilon. ■ Note: this is an experimental feature, and in future versions of *Mathematica* it may not be supported, or may have a different specification.

■ Experimental`$SameQTolerance

$SameQTolerance gives the number of decimal digits by which two numbers can disagree and still be considered the same according to SameQ.

The default setting is equal to Log[10, 2], corresponding to a tolerance of one binary digit. ■ See also: $MachineEpsilon. ■ Note: this is an experimental feature, and in future versions of *Mathematica* it may not be supported, or may have a different specification.

Index to Addendum